明清家具图集
2
（第二版）

上海大师室内设计研究所
康海飞　主编

中国建筑工业出版社
China Architecture & Building Press

图书在版编目（CIP）数据

明清家具图集2/康海飞主编.—2版.—北京：中国建筑工业出版社，2009

ISBN 978-7-112-11271-5

Ⅰ.明… Ⅱ.康… Ⅲ.家具-中国-明清时代-图集 Ⅳ.TS666.204-64

中国版本图书馆CIP数据核字（2009）第151641号

  中华民族在历史的长河中积累了辉煌的艺术文化，家具也是其艺术宝库中的重要组成部分，形成了独特风格的造型艺术。明清更是我国古代家具最灿烂的时期，将中国家具发展到了历史的顶峰，一直备受世人推崇和珍赏，业已成为全人类的文化遗产。

  本图集包括明清时期的椅、凳、桌、案、几、床榻、柜格、屏、台、架等精美家具，汇集了230件家具制作图、500个纹样图。每件家具都有它的名称（中英文对照）、立体造型图、雕花大样图、主视图、侧视图、俯视图以及各式各样的纹样图块。并附有光盘一张。图样精致，内容完整，是一部不可多得的图集。

  本书对家具生产、科学技术与艺术文化研究提供了宝贵的参考资料，可供国内外建筑设计师、室内设计师、家具设计师、工艺美术师、画家、雕塑家、大专院校师生及广大爱好者学习、欣赏、参考。

责任编辑：朱象清　费海玲
责任设计：崔兰萍
责任校对：刘　钰　陈晶晶

明清家具图集2（第二版）
上海大师室内设计研究所
康海飞　主编
＊
中国建筑工业出版社出版、发行（北京西郊百万庄）
各地新华书店、建筑书店经销
北京嘉泰利德公司制版
北京中科印刷有限公司印刷
＊
开本：889×1194毫米　1/16　印张：20　字数：640千字
2009年10月第二版　2015年4月第九次印刷
定价：**78.00**元（含光盘）
ISBN 978-7-112-11271-5
　　　（18444）

版权所有　翻印必究
如有印装质量问题，可寄本社退换
（邮政编码100037）

本 书 编 委 会

编委会主任： 康海飞

编委会顾问： 黄祖权（台湾）

编委会委员：（按姓氏笔画顺序）

王逢瑚　邓背阶　叶　喜　申黎明　关惠元　刘文金

李光耀　周　越　李克忠　吴智慧　宋魁彦　陈忠华

张亚池　张宏健　徐　雷　黄祖槐　薛文广　戴向东

编委会委员由同济大学、东北林业大学、南京林业大学、中南林业大学、北京林业大学、西南林业大学、中华建筑师事务所（台湾）等单位的教授组成，其中有博士生导师11位、硕士生导师4位。

主　　编： 康海飞

副 主 编： 石　珍

主编单位： 上海大师室内设计研究所

参编人员： 许志善　康国飞　黄　英　刘国庆　周锡宏　张　振

设计人员： 葛中华　竺雷杰　张蓓勤　王俊杰　康熙岳　马　涛

陈　涛　袁　博　郑冬毅　魏　娇　沈　骏　赵月姿

沈　岚　虞　佳　王　敏　陆　蓉　张　瑾　康　晶

陈伟俊　周　琦　胡美素　吴琦凤　夏彩峰　李嘉庚

陆晨平　崔　彬　易建军　施　颖　都嘉亮　柯向嘉

任亚辉　李晓希

编 者 的 话

《明清家具图集》2，自2007年出版以来，颇受国内外读者的欢迎，要求多次重印。同时，本书编委会在接待国内外读者的咨询中，发现原第一版在家具图的比例和结构方面存在一些问题。这次第二版对其作了大量改进，修改或调换了35%的图样，充实了更好的内容，使本书更加完美，更加实用。

被世人喻为中国文化艺术瑰宝的明清家具，内容丰富，形式多样，种类繁多。明代家具造型简练质朴，比例匀称，线脚丰富多彩；雕刻有浅浮雕、深浮雕、阳刻、阴刻等，且装饰花样千变万化。清代家具集历代精华于一朝，雕、刻、嵌、描、绘、堆漆、剔犀、镶金、饰件等，工艺高超精湛。明清家具常用纹饰有：龙纹、凤纹、麒麟纹、瑞兽纹、狮子纹、象纹、鹿纹、骏马纹、蟠螭纹、蜗纹、兽面纹、狮首衔环纹、蝉纹、绳纹、鱼纹、蝙蝠纹、云燕纹、莲花纹、梅花纹、牡丹纹、石榴纹、桃子纹、桃花纹、水仙花纹、菊花纹、竹子纹、灵芝纹、葫芦纹、葡萄纹、花鸟纹、海棠纹、柿蒂纹、缠枝纹、绞藤纹、花草纹、卷草纹、瓜蔓纹、花篮纹、云纹、回纹、百吉纹、如意纹、方胜纹、冰凌纹、火珠纹、八仙纹、汉字纹、山水纹、几何纹、圆环纹、拐子纹、卷云纹、博古纹、雷纹、勾卷纹、双钱纹、长寿纹、拜寿纹、五福捧寿纹、万福纹、长春纹、和合纹、连环纹、云鹤纹、耕读渔樵纹、人物故事纹、戏曲故事纹等。铜饰件的式样、种类繁多，有圆形、长方形、如意形、海棠形、环形、桃形、葫芦形、蝴蝶形、蝙蝠形等，明清家具图形千变万化，花纹繁缛纤细，因此必须精绘细画，其难度之大，功夫之深，是常人都可以想像到的。之所以数百年来未曾出版过一本完整的明清家具图集，也许与这些原因有关吧！

本书汇集了约230件明清家具的制作图、500个纹样图。每件家具都有它的名称、立体造型图、雕花大样图、主视图、侧视图、俯视图及详细尺寸。图样精致、内容完整，是难得的家具图集。

本书由康海飞编著并主持设计，得到各地编委的热情指导。因明清家具内容相当丰富涉及不同时期，不同地区，不同形式，所以编著与设计的难度相当大，由于我们专业水平有限，难免有错误和不足之处，希望国内外广大读者提出宝贵意见，我们将不胜感激。

本书附送的光盘包含了大部分常用家具的主视图、侧视图、俯视图的三视轮廓图。它既可作为创意设计时的参考，现成的图块又可直接借鉴、直接使用。既加快了家具设计制图的速度，又提高了创意设计的水平，也可以指导家具加工，实用性强，为明清家具设计、制图、生产加工提供了方便快捷的工具。书中的图形均编页码，读者可查询本书所附光盘中相应的页码文件名，用CAD2004以上版本打开文件，即可取得相应图块，附书光盘必须与本书配合才能使用。

明清家具种类繁多，形式多样，本书因篇幅有限，难以全部收进。如有读者需要另外家具的施工图纸，可与本书编委会取得联系。联系地址：上海市共和新路425号凯鹏国际大厦21楼，邮编：200070，咨询电话：021-56310018　传真：021-56318918。

《明清家具图集》编委会

前　言

　　中国是世界上最古老又有着悠久文化的国家。中华民族在历史的进程中积累了众多辉煌的艺术文化，家具也是其艺术宝库中的重要组成部分，它历经能工巧匠的智慧与创造，逐渐形成了具有独特风格的造型艺术。

　　明代和清代，是中国古代家具最辉煌灿烂的时期。在明清家具中，出类拔萃的是以花梨木、紫檀木、溪敕木、铁力木、香红木、楗木等为主要用材的优质硬木家具。又有京式、苏式、宁式、广式等风格之区别。明清家具将中国古代家具发展到了历史顶峰，几乎到了无以复加的地步。

　　明式家具和清式家具，以其鲜明的艺术风格，长期以来一直备受世人的推崇和珍赏。特别是明式家具精邃的文化内涵，体现了中华民族传统艺术的精华，也是世界文化宝库中不可多得的艺术瑰宝，业已成为全人类的优秀文化遗产。

　　我们编著本书的目的，旨在向国内外广大读者提供一份珍贵的文化遗产，希望大家研究它、学习它，把一切优秀的传统表现手法和艺术特点作为现代建筑设计、艺术设计和工业产品造型设计的有益借鉴。

　　我国是世界上的家具生产大国，近年来，中国家具年产值超千亿元，但其中的出口份额仅占15％左右，与西方发达国家的出口比率恰好相反，原因是我国生产的家具缺乏自己的民族文化特征。当务之急是除了必须开发新材料之外，更应努力培养一批高级家具设计师。我们编著本书的目的是希望在这方面发挥一定的作用。

　　本书除了将对指导家具生产、科学技术和艺术文化历史研究提供可贵的参考资料外，还可供世界各国建筑设计师、室内设计师、家具设计师、工艺美术设计师、画家、雕塑家、大专院校师生和技术工人及业余爱好者参考。

　　今天，《明清家具图集2（第二版）》虽已出版，但是我们喜中有忧，因为挂一漏万，尚有许多内容没有编进本书中，且部分图样有不足之处，这些问题有待于我们再版时充实提高，希望广大读者提出批评指正。

目 录

椅 类 Chairs
- 007 宝　座　Thrones
- 010 扶手椅　Armchairs
- 035 靠背椅　Side chairs
- 038 玫瑰椅　Rose chairs
- 039 官帽椅　Official's hat armchairs
- 041 交　椅　Folding armchairs
- 042 太师椅　Fauteuils
- 044 圈　椅　Round-backed armchairs
- 048 矮坐椅　Low seat armchairs
- 050 屏背椅　Screen-backed chairs
- 051 高背文椅　Writing chairs with high back

凳 类 Stools
- 052 春　凳　Large benches
- 053 脚　凳　Foot stools
- 054 海棠式凳　Begonia-shaped stools
- 055 文竹凳　Stools with setose asparagus motif
- 056 坐　墩　Seat stools
- 061 小条凳　Small narrow benches
- 063 方　凳　Square stools
- 071 方　机　Square stools
- 072 扇形凳　Fan-shaped stools
- 073 圆　凳　Round stools

桌 类 Tables
- 078 方　桌　Square tables
- 088 长　桌　Rectangular tables
- 097 半　桌　Half tables
- 101 琴　桌　Lute tables
- 103 书　桌　Recessed-legs tables with drawers
- 104 供　桌　Altar tables
- 106 条　桌　Narrow rectangular tables with corner legs
- 112 炕　桌　Kang tables
- 117 房前桌　Tables in bedroom put in front of windows
- 119 鱼　桌　Fish tables
- 120 大圆台　Large round tables
- 121 圆　桌　Round tables
- 122 麻将桌　Mah-jong tables

案 类 Recessed-leg tables
- 123 画　案　Painting tables
- 125 炕　案　Recessed-leg Kang tables
- 126 翘头案　Recessed-legs tables with everted flanges on the top
- 134 平头案　Flat-top narrow recessed-leg tables
- 138 卷头案　Narrow recessed-leg tables with scroll termination
- 139 条　案　Narrow recessed-leg tables

几 类 Stands and tables
- 141 香　几　Incense or plant stands
- 157 花　几　Flower stands
- 161 茶　几　Tea tables
- 166 方　几　Square tables
- 168 炕　几　Narrow *kang* tables
- 173 长　几　Long rectangular tables
- 174 下　卷　Narrow rectangular tables with open-work carving on solid board legs

床 榻 类 Beds
- 175 架子床　Canopy beds
- 192 罗汉床　Luohan beds
- 202 凉　床　Cool beds
- 203 美人榻　Daybeds for beautiful women

柜 格 类 Cabinets
- 205 衣　柜　Wardrobes
- 208 联二橱　Two-drawer coffers
- 214 床头柜　Bedstands
- 215 矮　柜　Low cabinets
- 216 柜　格　Display cabinets
- 225 书　格　Book shelves
- 226 多宝格　Display cabinets or shelves

其 他 类 Others
- 237 屏　类　Screens
- 251 台　类　Stands and platforms
- 262 架　类　Stands and racks
- 270 箱　类　Boxes and cases
- 273 佛　龛　Buddha chests

附 录 Appendixes
- 274 木雕书法　Calligraphy designs of wood carving
- 275 构件纹样　Designs and motifs
- 298 雕花图块　Carving pattern block
- 317 构件名称　Components

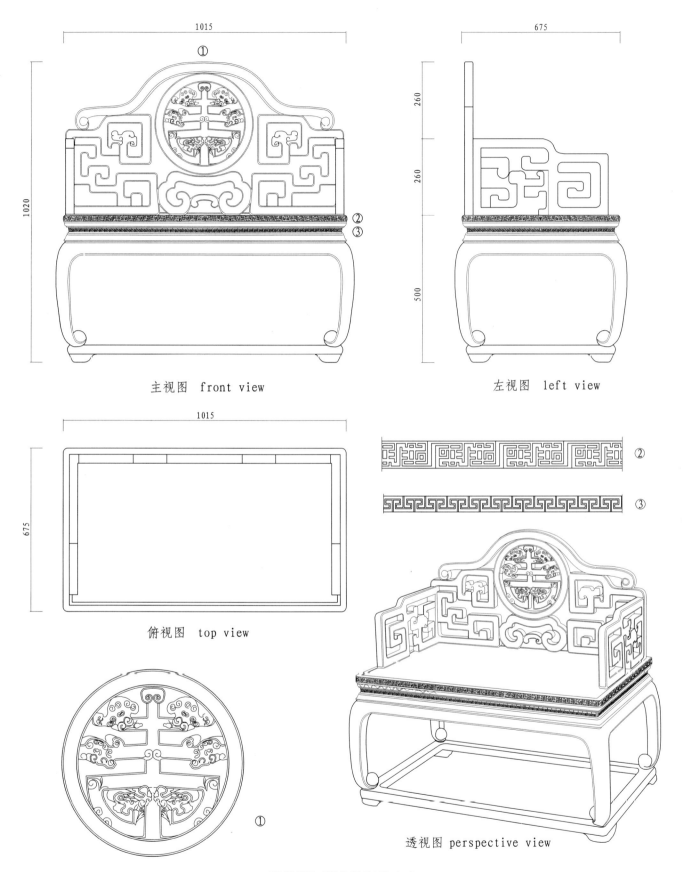

明代剔红夔龙捧寿纹宝座
Ming dynasty carved cinnabar lacquer throne with *kui*-dragon holding a *shou* character motif

主视图 front view

左视图 left view

俯视图 top view

透视图 perspective view

清中期紫檀小宝座
Mid Qing dynasty *zitan* wood small throne

宝座 9

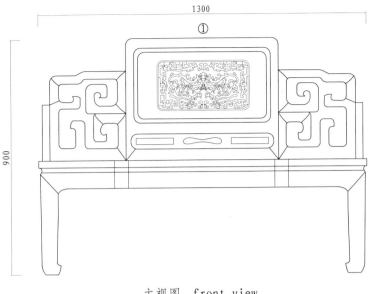

主视图 front view

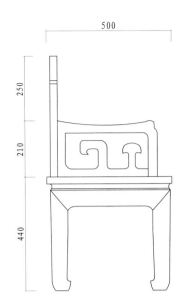

左视图 left view

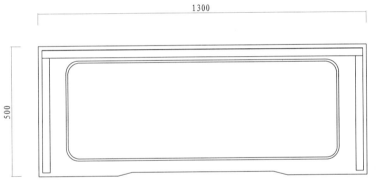

俯视图 top view

透视图 perspective view

清代红木宝座
Qing dynasty *hong* wood throne

清代酸枝镶黄铜花纹大理石凹栳扶手椅
Qing dynasty *suanzhi* wood armchair with brass and flower pattern marble inlay

清乾隆紫檀西洋花纹扶手椅
Qing dynasty Qianlong period *zitan* wood armchairs with western pattern

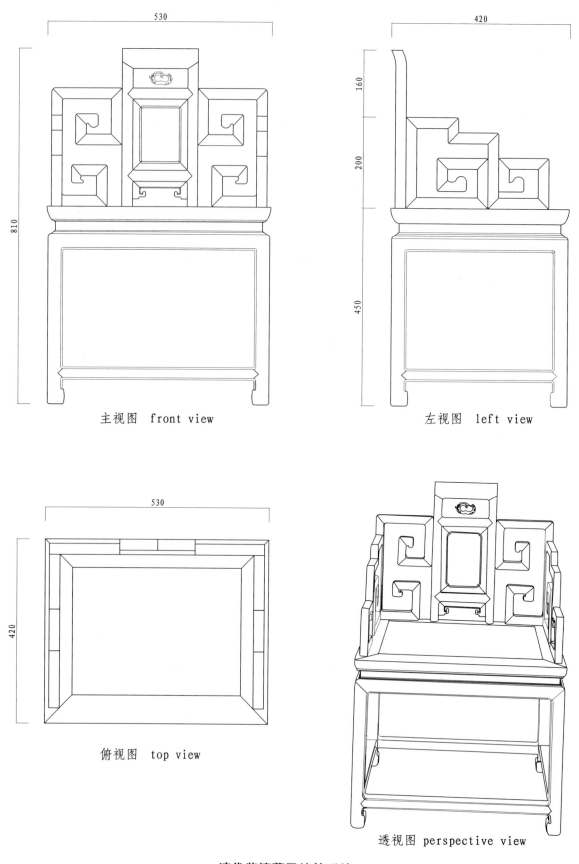

清代紫檀夔凤纹扶手椅
Qing dynasty zitan wood armchair with kui-phoenix motif

主视图 front view

左视图 left view

俯视图 top view

透视图 perspective view

清代紫檀福寿纹扶手椅
Qing dynasty *zitan* wood armchair with *fu-shou* character motif

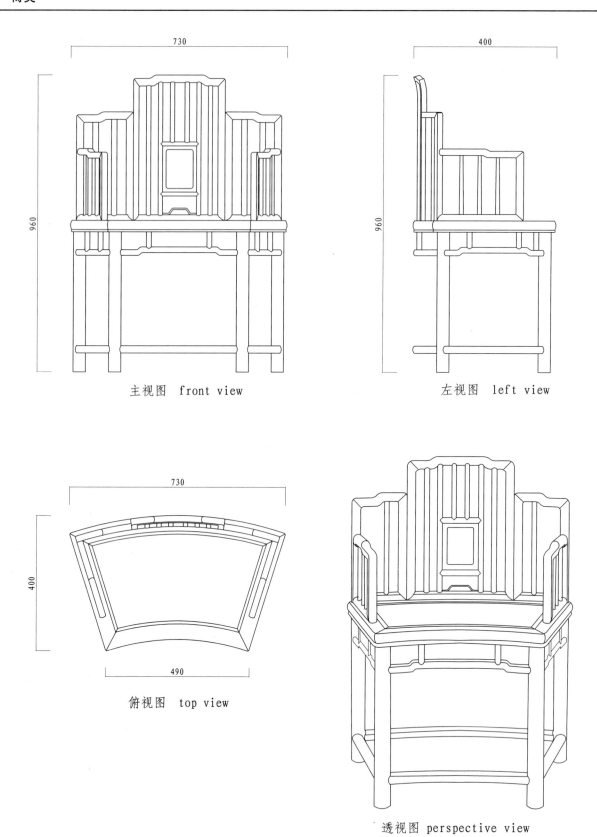

主视图 front view

左视图 left view

俯视图 top view

透视图 perspective view

清中期铁力木扇形坐面扶手椅
Mid Qing dynasty *tieli* wood armchair with fan-shaped seat

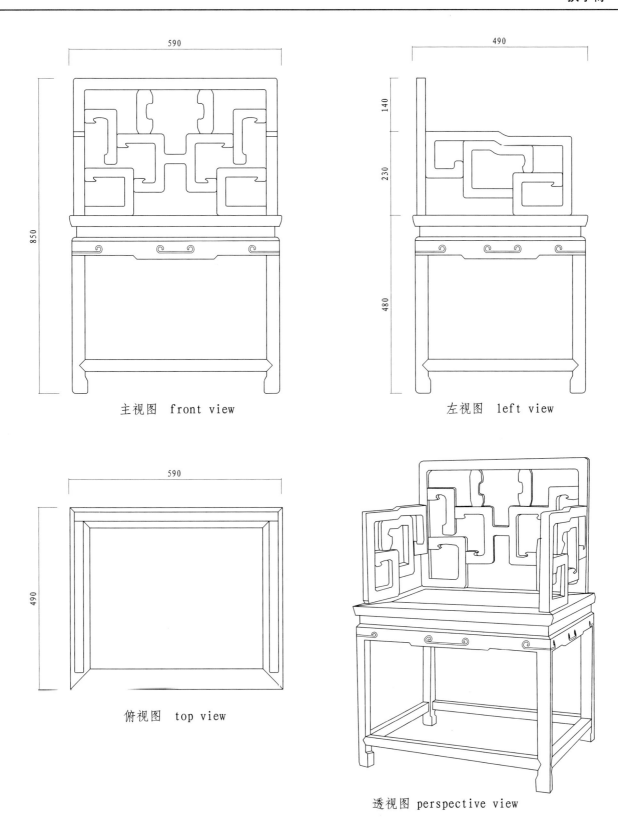

明代红木勾云纹短背扶手椅
Ming dynasty *hong* wood armchair with low back and hook-cloud design

16　椅类

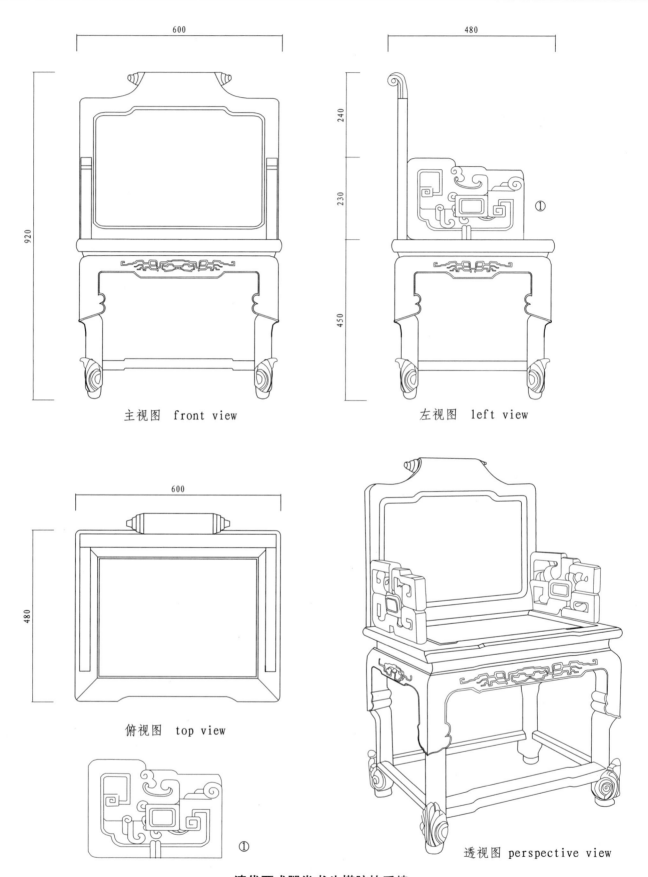

主视图 front view

左视图 left view

俯视图 top view

透视图 perspective view

清代两式腿卷书头搭脑扶手椅
Qing dynasty armchair with double-typed legs and cylindrical scroll on top rail

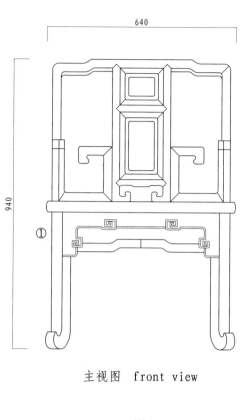

主视图 front view

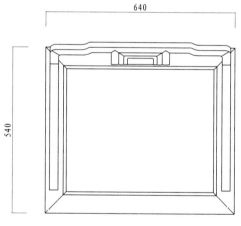

左视图 left view

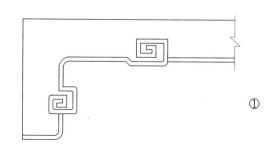

俯视图 top view

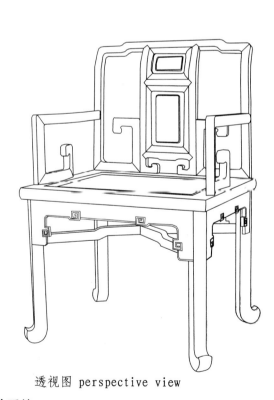

透视图 perspective view

清中期鸡翅木勾云纹扶手椅
Mid Qind dynasty *jichi* wood armchair with hook-cloud design

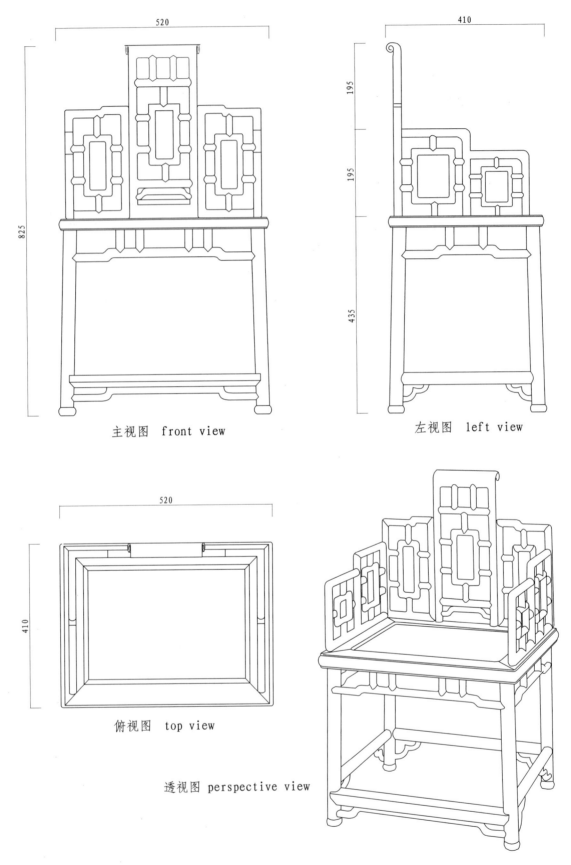

主视图 front view

左视图 left view

俯视图 top view

透视图 perspective view

清早期乌木七屏卷书式扶手椅

Early Qing dynasty ebony armchair with seven screens and cylindrical scroll

扶手椅 19

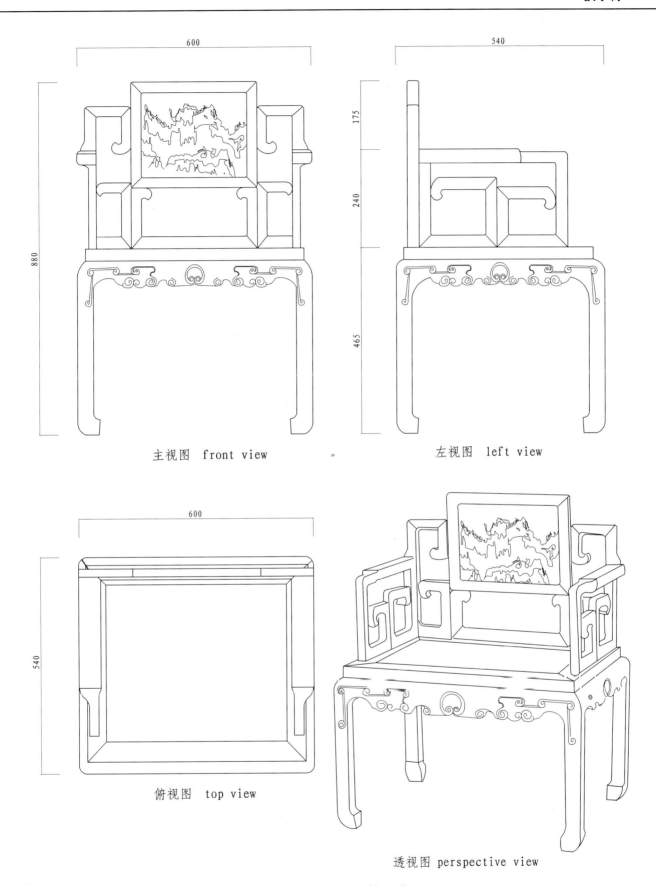

主视图 front view

左视图 left view

俯视图 top view

透视图 perspective view

清代酸枝镶大理石扶手椅
Qing dynasty *suanzhi* wood armchair with marble inlay

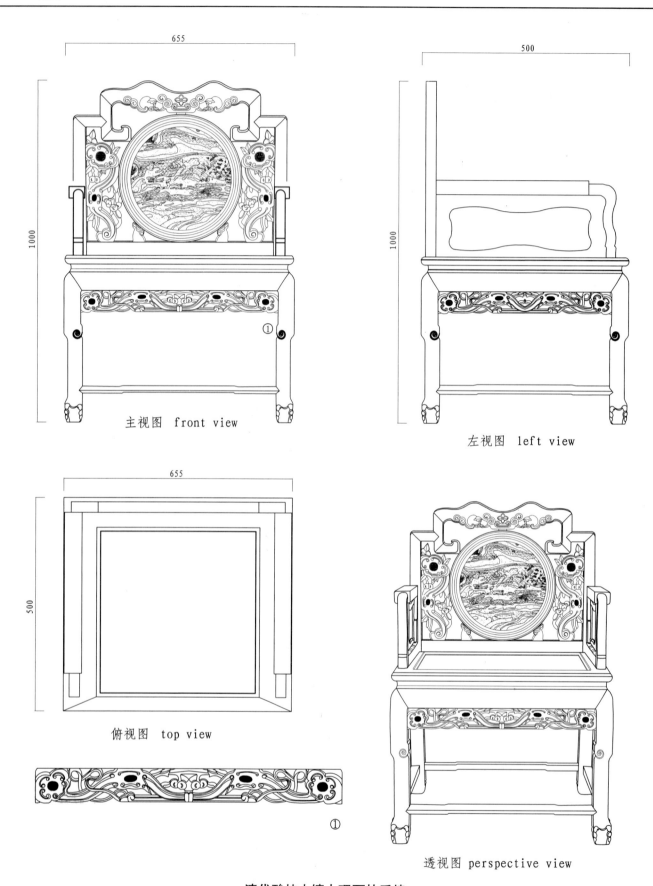

清代酸枝木镶大理石扶手椅
Qing dynasty *suanzhi* wood armchair with marble inlay

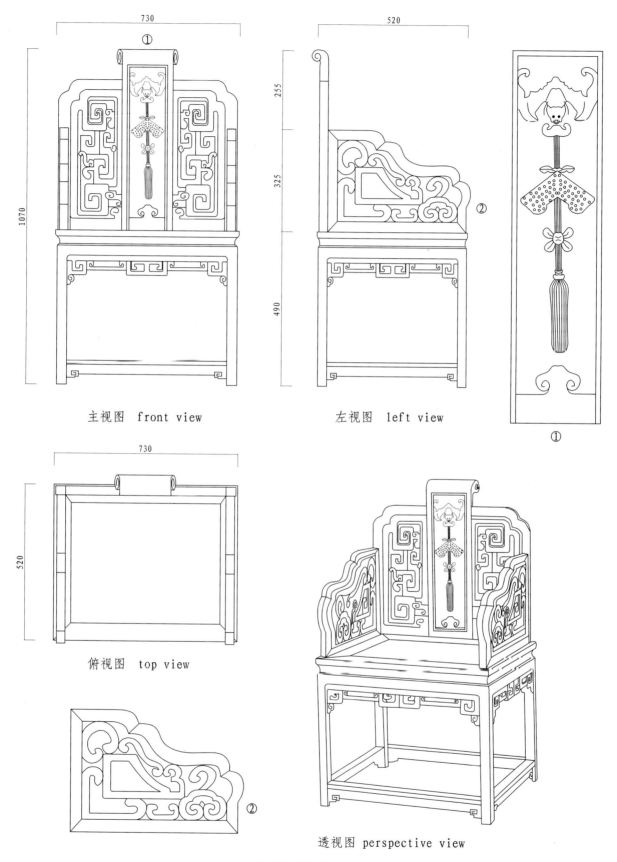

主视图 front view

左视图 left view

俯视图 top view

透视图 perspective view

清中期紫檀扶手椅
Mid Qing dynasty *zitan* wood armchair

主视图 front view

左视图 left view

俯视图 top view

透视图 perspective view

清代红木镶大理石屏背扶手椅
Qing dynasty *hong* wood armchair with screen back and marble inlay

主视图 front view

左视图 left view

俯视图 top view

透视图 perspective view

清代红木短背勾云纹扶手椅
Qing dynasty *hong* wood armchair with low back and hook-cloud motif

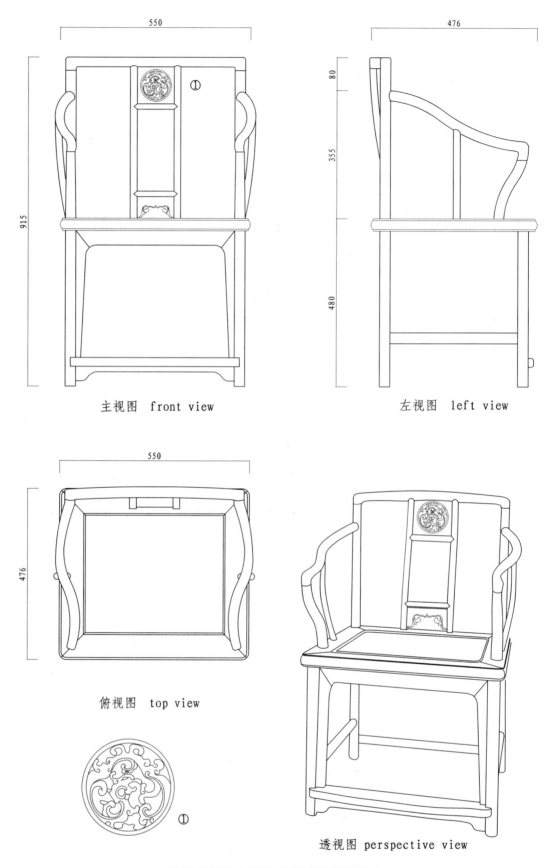

主视图 front view
左视图 left view
俯视图 top view
透视图 perspective view

清代老花梨木高扶手带联帮棍矮背椅
Qing dynasty old *huali* wood armchair with high arms, side posts and low back

清乾隆紫檀嵌黄杨木蝠螭纹扶手椅
Qing dynasty Qianlong period *zitan* wood armchair with boxwood inlay and bat-and-hornless-dragon pattern

椅类

主视图 front view

左视图 left view

俯视图 top view

透视图 perspective view

清代紫檀云蝠纹扶手椅
Qing dynasty *zitan* wood armchair with cloud-and-bat pattern

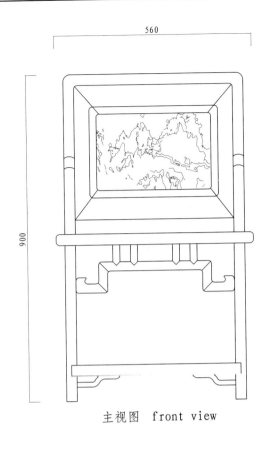

主视图 front view

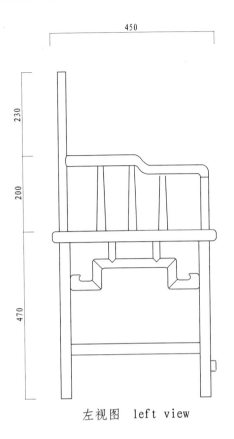

左视图 left view

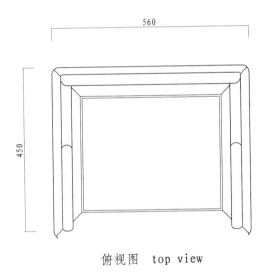

俯视图 top view

透视图 perspective view

清代红木镶大理石屏背扶手椅
Qing dynasty *hong* wood armchair with marble inlay and screen back

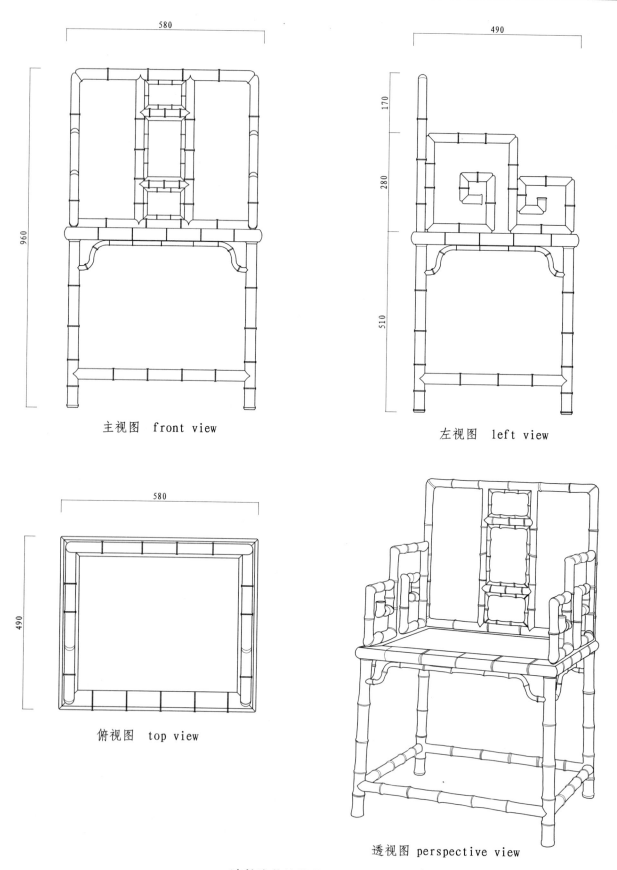

主视图 front view

左视图 left view

俯视图 top view

透视图 perspective view

清乾隆紫檀竹节纹扶手椅
Qing dynasty Qianlong period *zitan* wood bamboo-shaped armchair

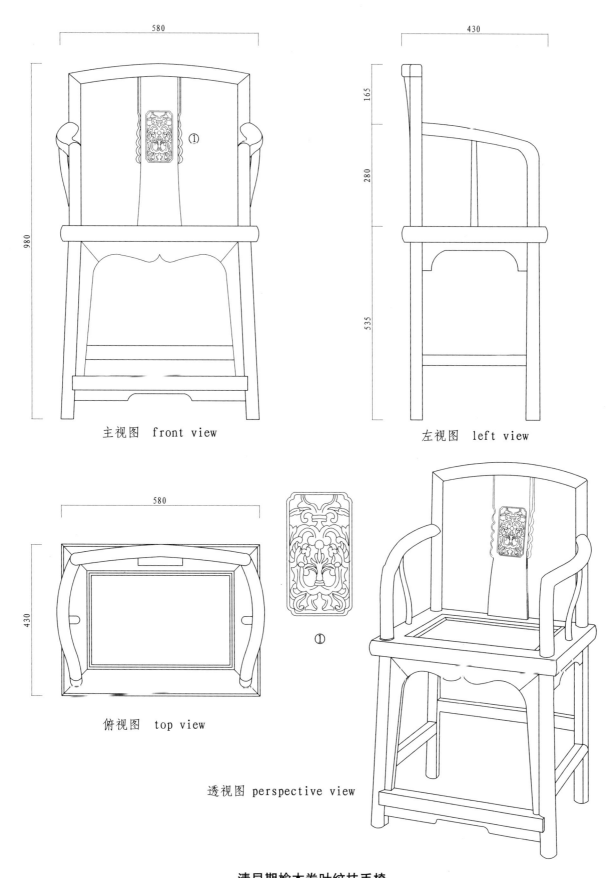

主视图 front view

左视图 left view

俯视图 top view

透视图 perspective view

清早期榆木卷叶纹扶手椅
Early Qing dynasty elm wood armchair with curling tendril design

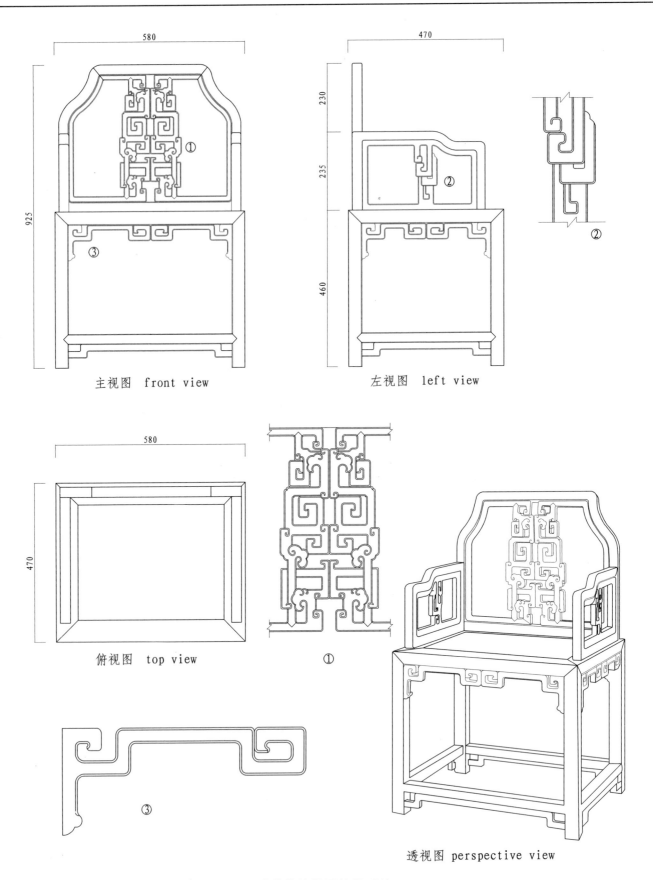

清代紫檀拐子纹扶手椅
Qing dynasty *zitan* wood armchair with rectangular spiral pattern

主视图 front view

左视图 left view

俯视图 top view

透视图 perspective view

清中期黄花梨扶手椅
Mid Qing dynasty *huanghuali* wood armchair

32 椅类

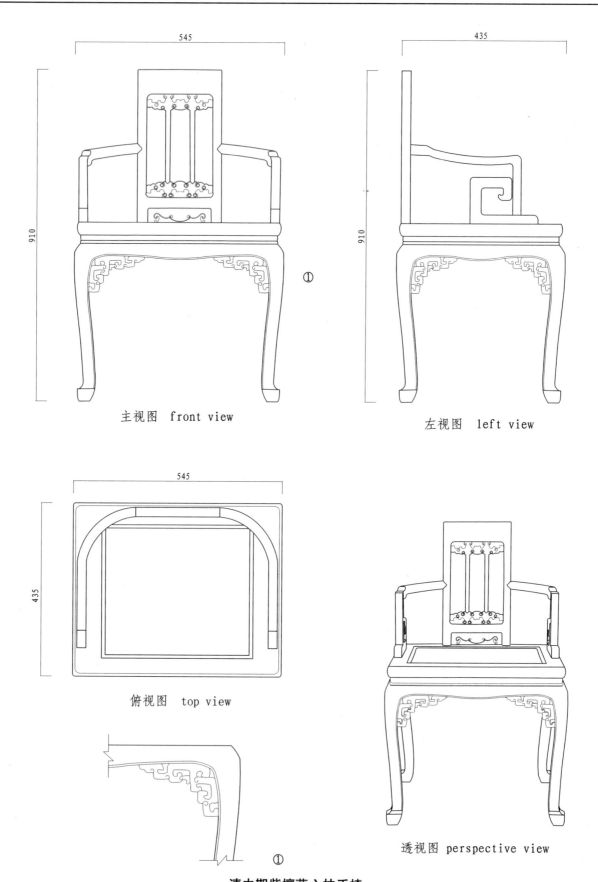

主视图 front view

左视图 left view

俯视图 top view

透视图 perspective view

清中期紫檀藤心扶手椅
Mid Qing dynasty *zitan* wood armchair with rattan seat

主视图 front view

左视图 left view

俯视图 top view

透视图 perspective view

清中期紫檀嵌珐琅扶手椅
Mid Qing dynasty *zitan* wood armchair with enamel inlay

主视图 front view

左视图 left view

俯视图 top view

透视图 perspective view

清代酸枝木延年款扶手椅
Qing dynasty *suanzhi* wood armchair with *yannian* character design

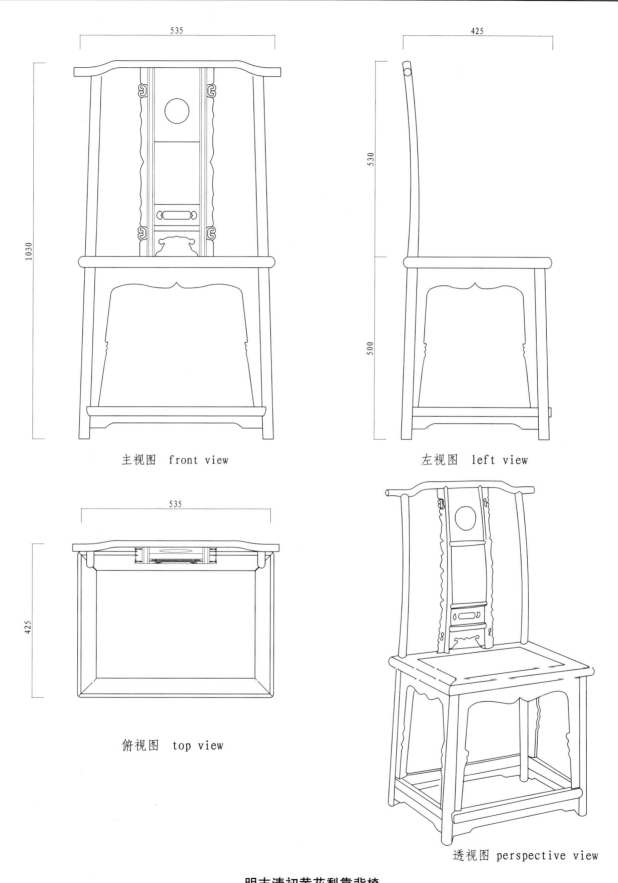

主视图 front view

左视图 left view

俯视图 top view

透视图 perspective view

明末清初黄花梨靠背椅
Early Qing dynasty *huanghuali* wood side chair

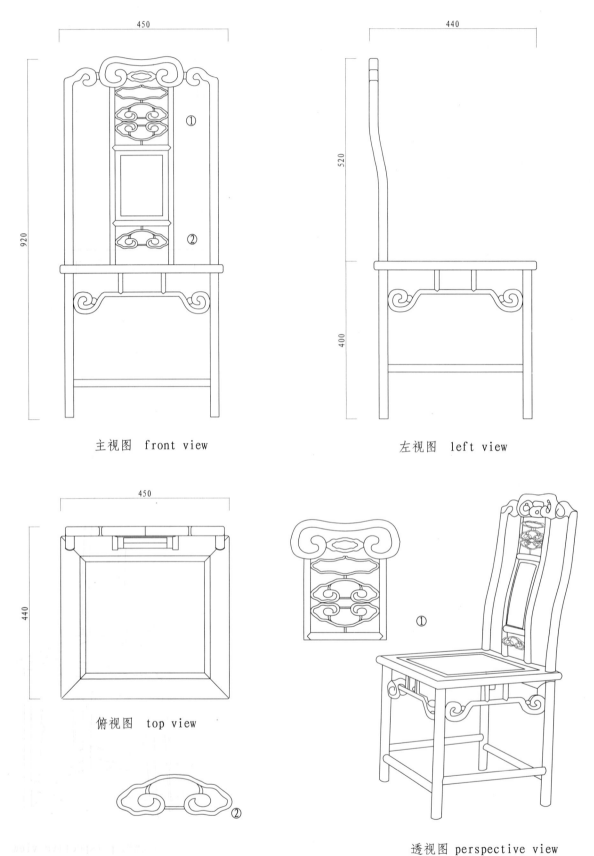

主视图 front view

左视图 left view

俯视图 top view

透视图 perspective view

清代黄花梨木如意头纹单靠背椅
Qing dynasty *huanghuali* wood side chair with *ruyi*-head motif

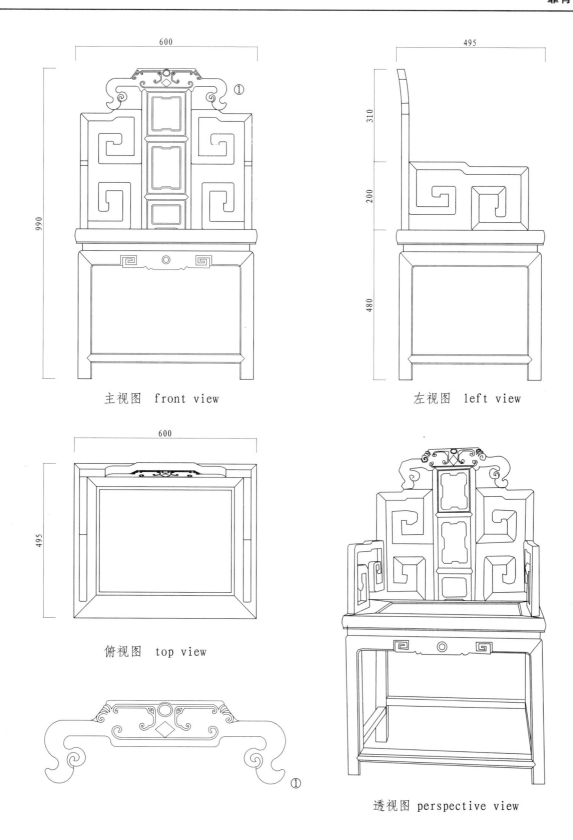

主视图 front view

左视图 left view

俯视图 top view

透视图 perspective view

清代核桃木硬鼓纹靠背椅
Qing dynasty walnut armchair with hard-drum pattern

38 椅类

主视图 front view

左视图 left view

俯视图 top view

透视图 perspective view

明代杞梓木双座玫瑰椅
Qing dynasty *qizi* wood double seat rose chair

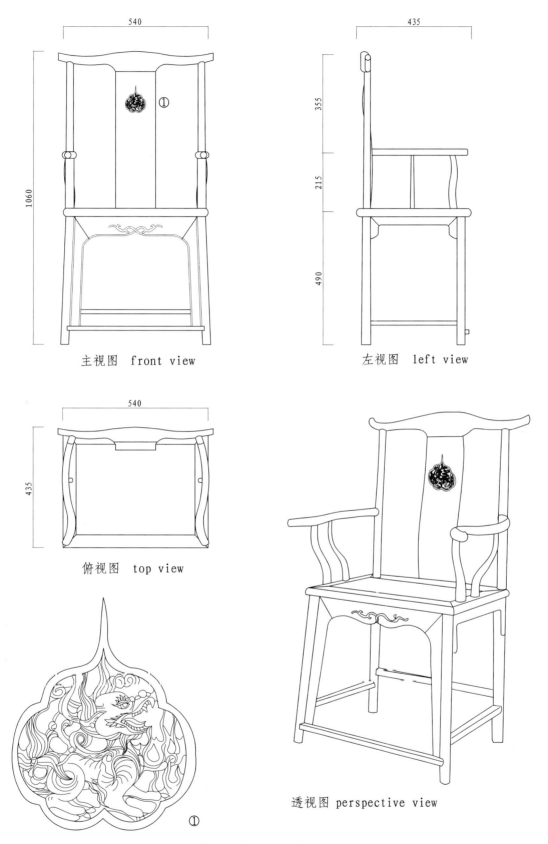

主视图 front view

左视图 left view

俯视图 top view

透视图 perspective view

明代红木瑞兽纹四出头官帽椅
Ming dynasty *hong* wood official's hat armchair with four protruding ends and lucky animals pattern

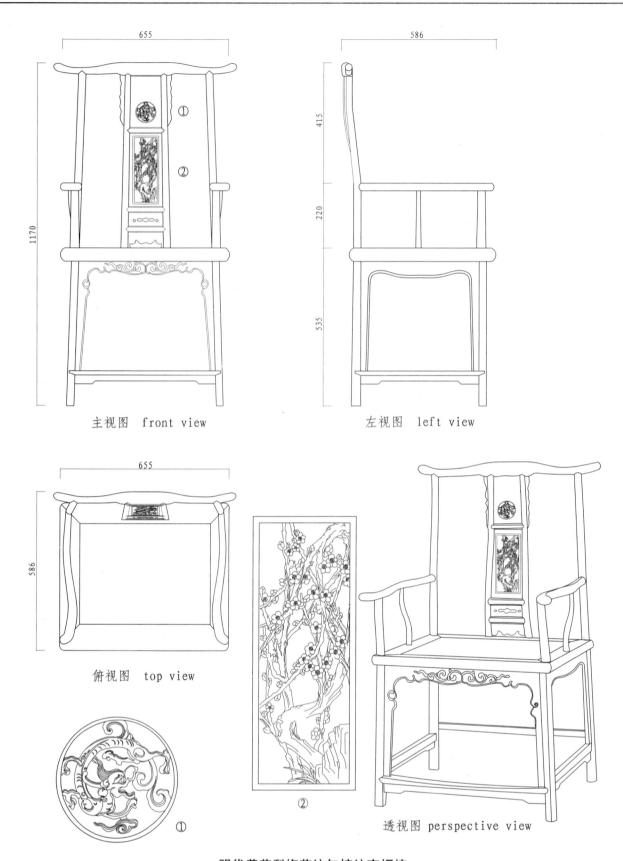

主视图 front view 左视图 left view 俯视图 top view 透视图 perspective view

明代黄花梨梅花纹与螭纹官帽椅
Ming dynasty *huanghuali* wood offical's hat armchair with plum blossom pattern and stylized hornless dragon design

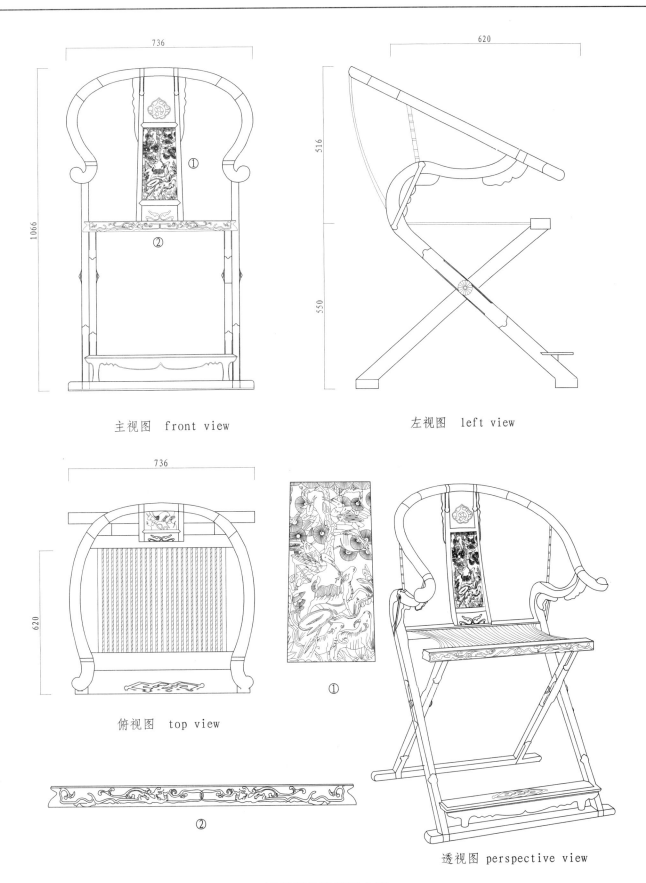

主视图 front view

左视图 left view

俯视图 top view

透视图 perspective view

清代黄花梨镶铜交椅
Qing dynasty *huanghuali* wood folding armchair inlaid with brass

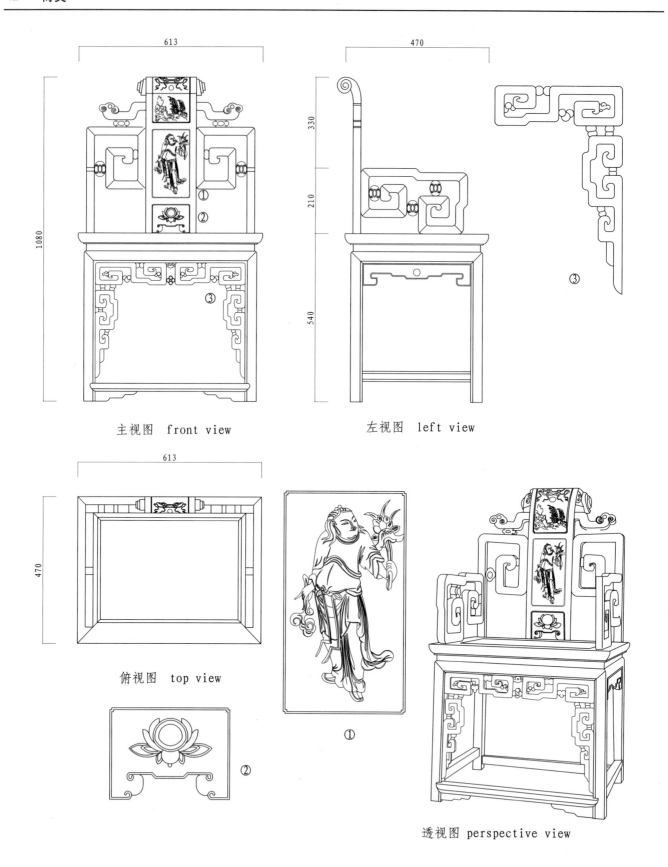

主视图 front view 左视图 left view 俯视图 top view 透视图 perspective view

清代榉木太师椅
Qing dynasty *ju* wood fauteuil

主视图 front view

左视图 left view

俯视图 top view

透视图 perspective view

清代红木大理石圆景太师椅
Qing dynasty *hong* wood fauteuil with round marble inlay

主视图 front view

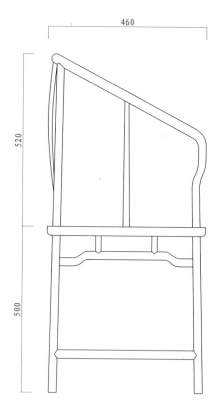
左视图 left view

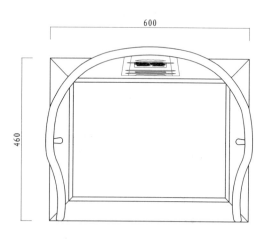
俯视图 top view

透视图 perspective view

清代黄花梨圈椅
Qing dynasty *huanghuali* wood round-backed armchair

圈椅 45

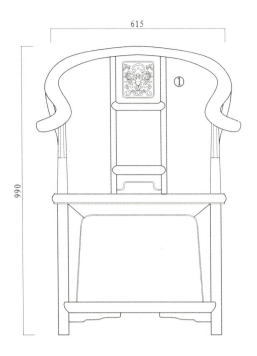

主视图 front view

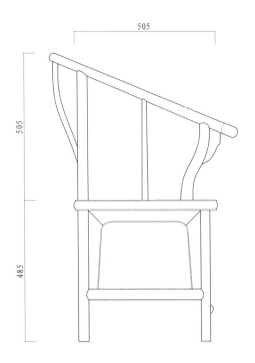

左视图 left view

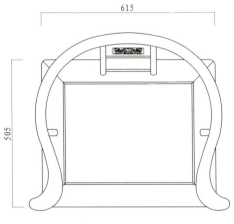

俯视图 top view

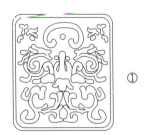

透视图 perspective view

清代鸡翅木圈椅
Qing dynasty *jichi* wood round-backed armchair

46　椅类

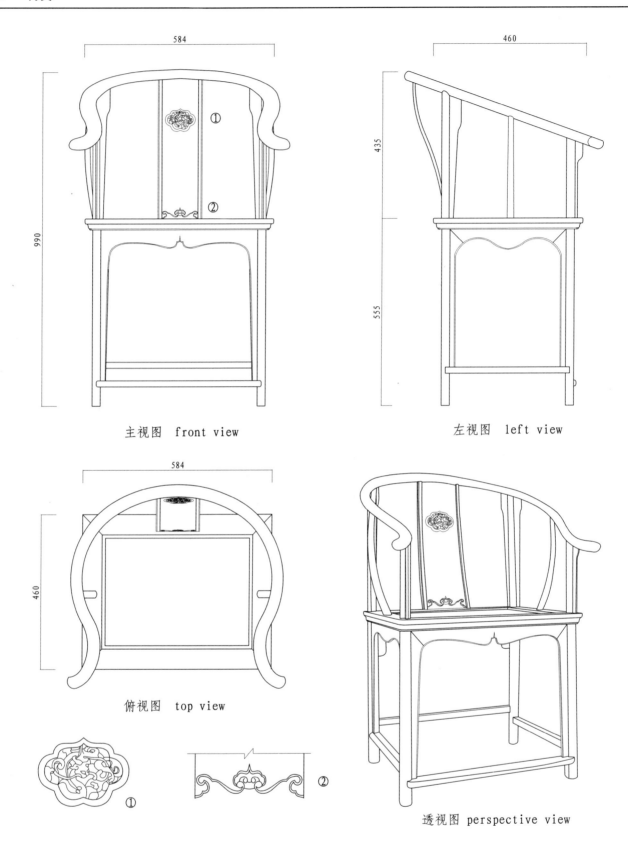

主视图　front view

左视图　left view

俯视图　top view

透视图　perspective view

清代老花梨木圈椅
Qing dynasty old *huali* wood round-backed armchair

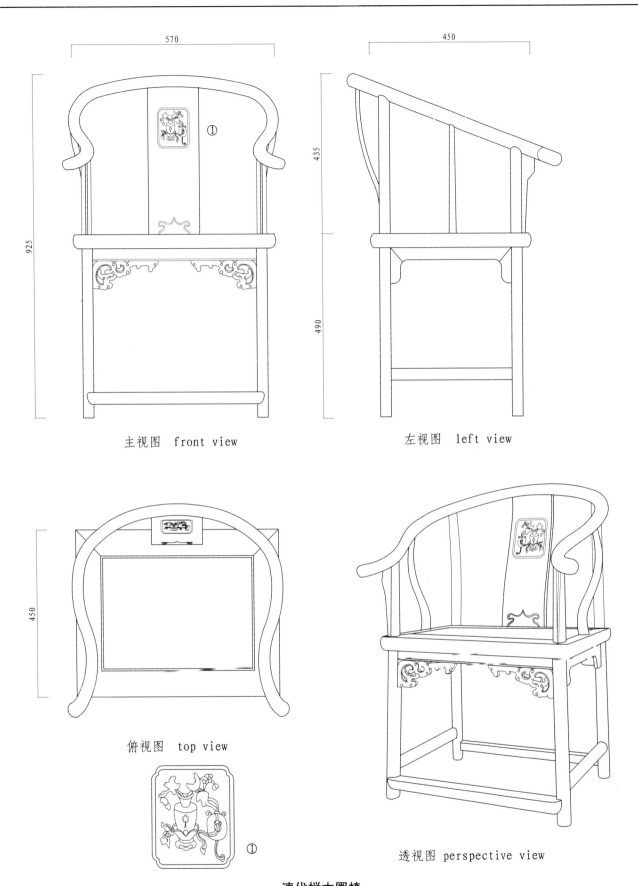

主视图 front view　左视图 left view　俯视图 top view　透视图 perspective view

清代榉木圈椅
Qing dynasty *ju* wood round-backed armchair

清早期黄花梨福寿纹矮坐椅

Early Qing dynasty *huanghuali* wood low seat armchair with *fu-shou* character design

锉坐椅 49

主视图 front view

左视图 left view

俯视图 top view

透视图 perspective view

清代酸枝木（密地屏）坐椅
Qing dynasty *suanzhi* wood armchair

主视图 front view

左视图 left view

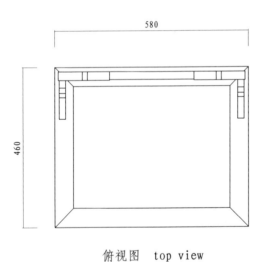

俯视图 top view

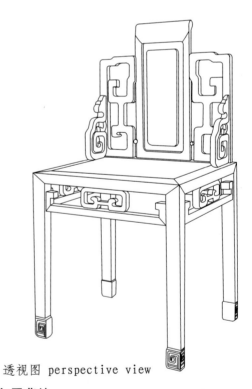

透视图 perspective view

清代红木三屏风式插角屏背椅
Qing dynasty *hong* wood screen-backed chair with three-panel screens and inserted corners

高背文椅 51

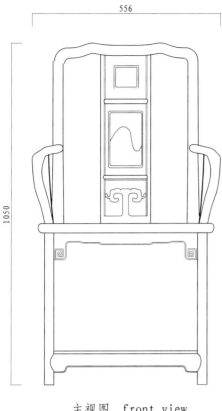

主视图 front view

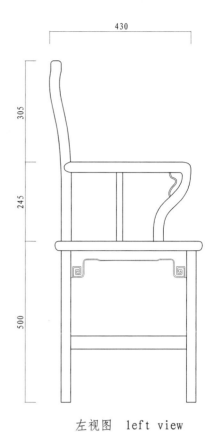

左视图 left view

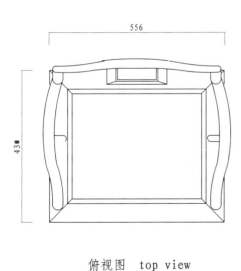

俯视图 top view

透视图 perspective view

清代红木镶云石高背文椅
Qing dynasty *hong* wood writing chair with high back
and cloud pattern marble inlay

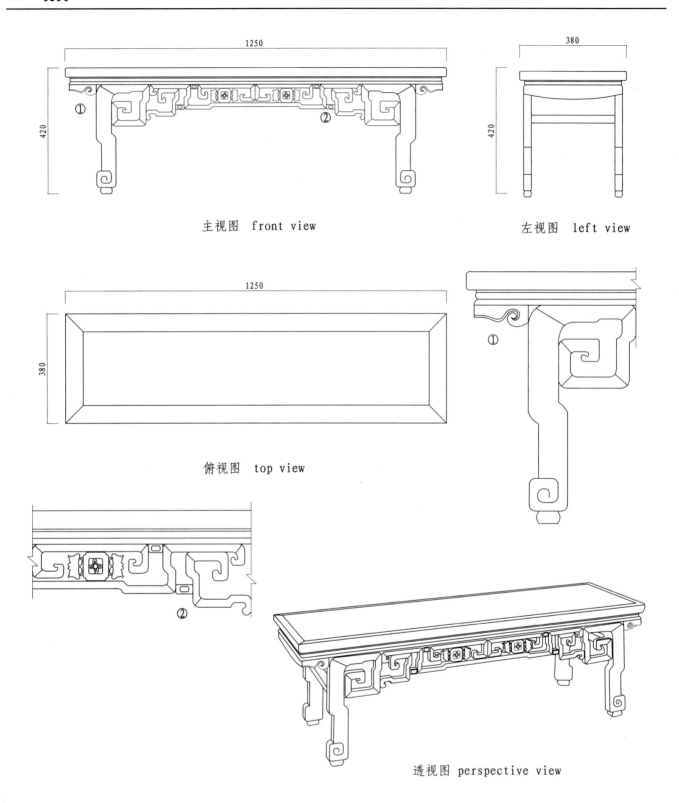

主视图 front view

左视图 left view

俯视图 top view

透视图 perspective view

清代榉木拐子纹春凳
Qing dynasty *ju* wood large bench with rectangular spiral pattern

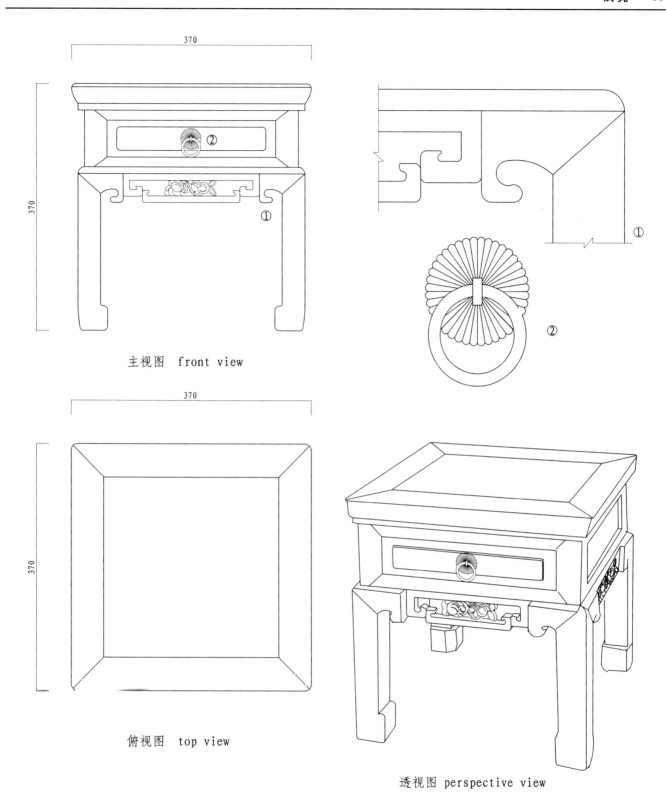

主视图 front view

俯视图 top view

透视图 perspective view

清代宁式榉木脚凳
Qing dynasty *Ningbo* style *ju* wood foot stool

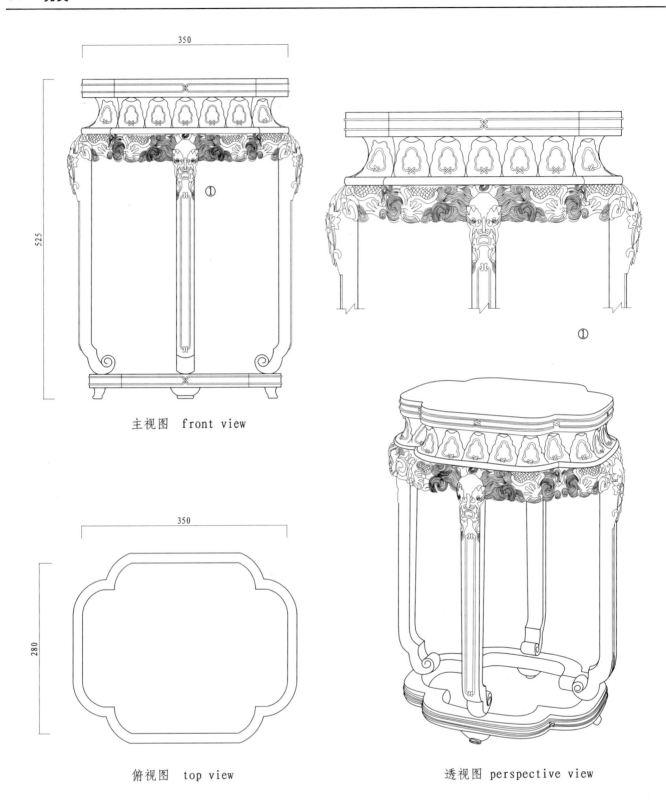

主视图 front view

俯视图 top view

透视图 perspective view

清乾隆云纹海棠式凳
Qing dynasty Qianlong period begonia-shaped stool with cloud design

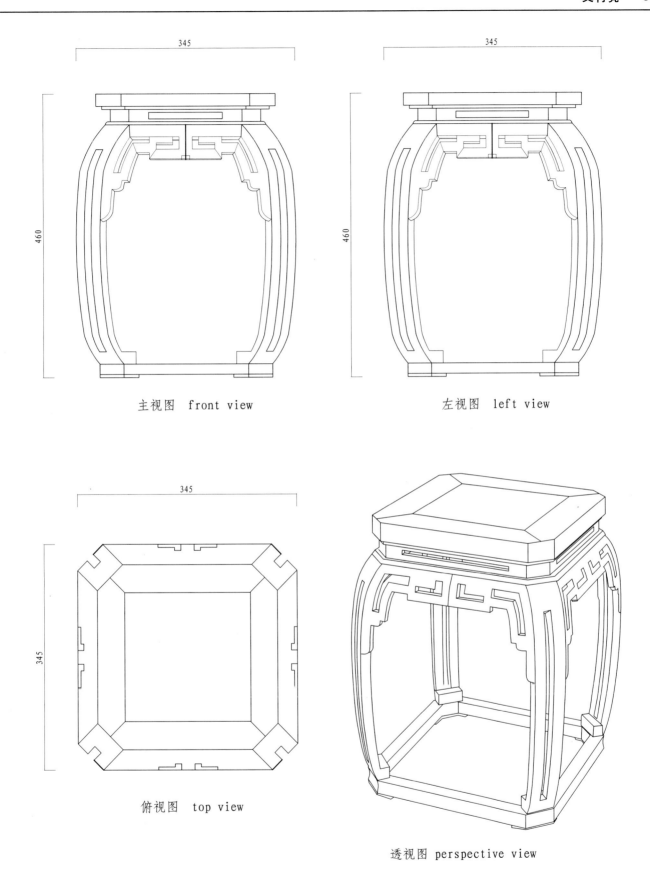

主视图 front view

左视图 left view

俯视图 top view

透视图 perspective view

清中期方形抹脚文竹凳
Mid Qing dynasty stool with setose asparagus motif and square frameworks with mortises

56 凳类

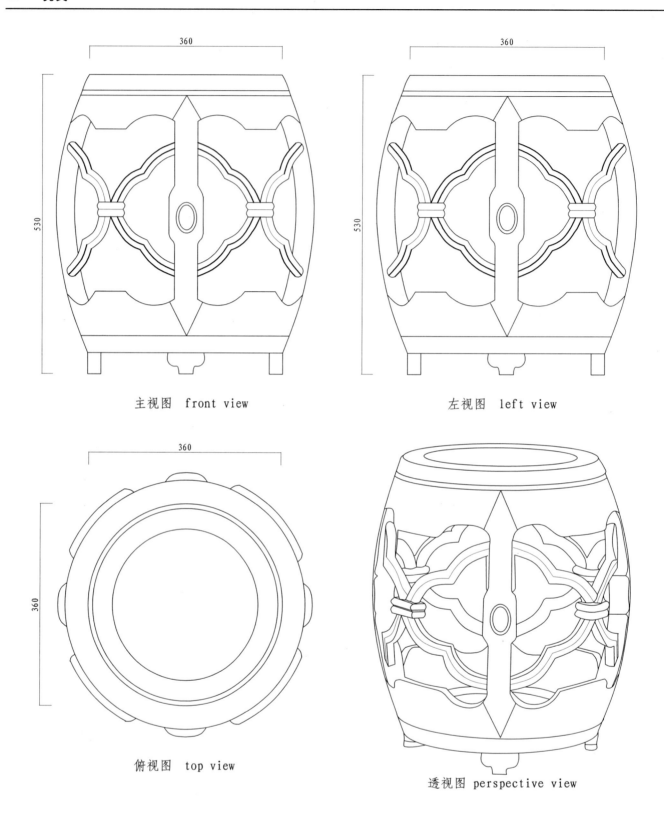

主视图 front view

左视图 left view

俯视图 top view

透视图 perspective view

清代红木四面开光坐墩
Qing dynasty *hong* wood drum stool with four openings

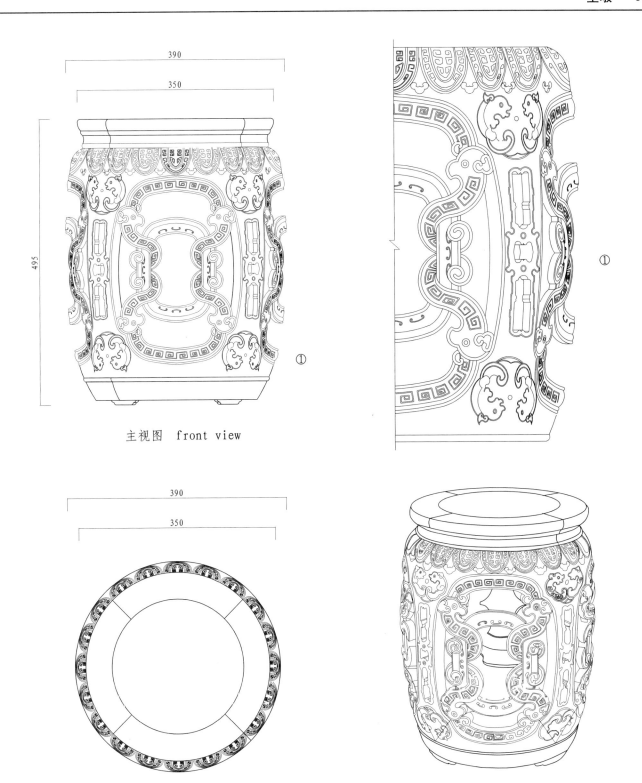

主视图 front view

俯视图 top view

透视图 perspective view

清代紫檀夔凤纹四开光坐墩
Qing dynasty *zitan* wood drum stool with four openings and *kui*-phoenix motif

主视图 front view

俯视图 top view

透视图 perspective view

清乾隆紫檀兽面衔环纹八方坐墩
Qing dynasty Qianlong period *zitan* wood octagonal stool with animal-mask holding a ring design

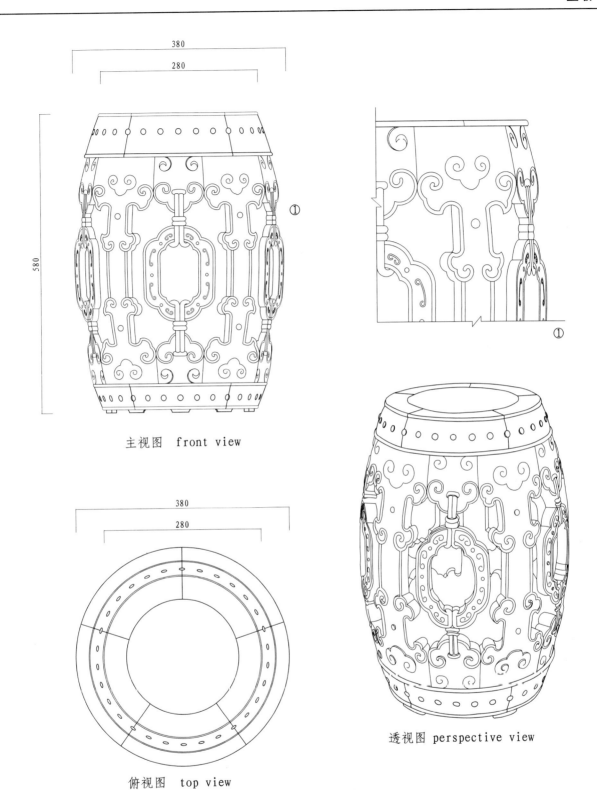

清乾隆紫檀云头纹五开光坐墩
Qing dynasty Qianlong period *zitan* wood drum stool with five openings and cloud motif

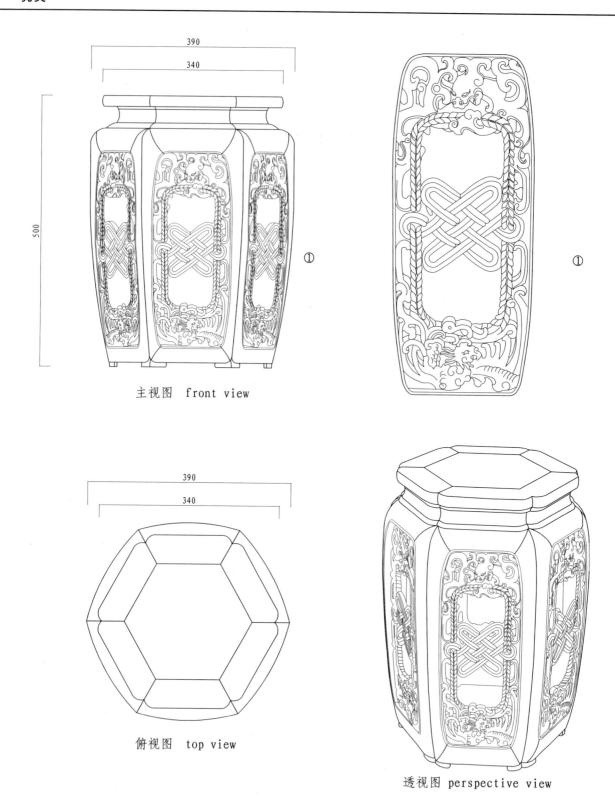

主视图 front view

俯视图 top view

透视图 perspective view

清乾隆紫檀龙凤盘肠纹六方坐墩

Qing dynasty Qianlong period *zitan* wood hexagonal stool with dragon, phoenix and twist ribbons motif

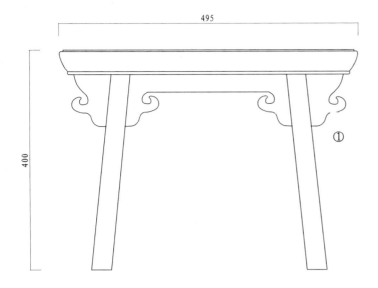

主视图 front view

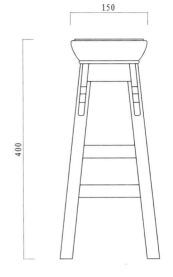

左视图 left view

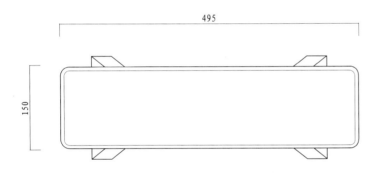

俯视图 top view

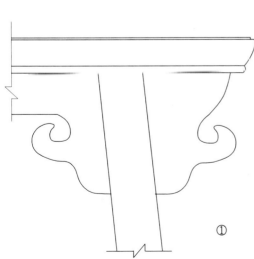

透视图 perspective view

清代榉木夹头榫小条凳
Qing dynasty *ju* wood small narrow bench with elongated bridle joints

主视图　front view

左视图　left view

俯视图　top view

透视图　perspective view

明代黄花梨条凳
Ming dynasty *huanghuali* wood narrow bench

主视图 front view

俯视图 top view

透视图 perspective view

清代红木嵌瓷板方凳
Qing dynasty *hong* wood square stool with porcelain inlay

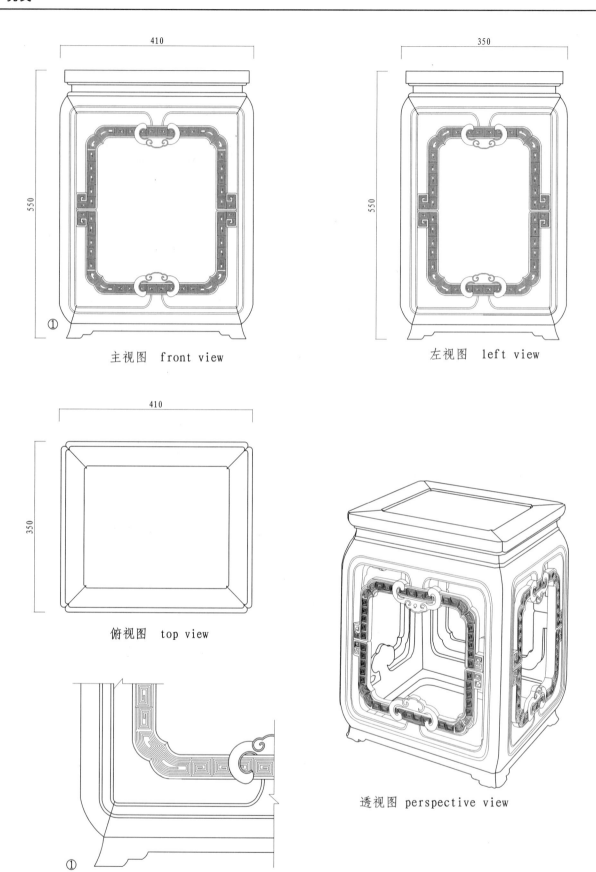

主视图 front view

左视图 left view

俯视图 top view

透视图 perspective view

清中期紫檀如意纹方凳
Mid Qing dynasty *zitan* wood square stool with *ruyi* motif

主视图 front view

俯视图 top view

透视图 perspective view

清代紫檀如意云头纹方凳
Qing dynasty *zitan* wood square stool with *ruyi* motif

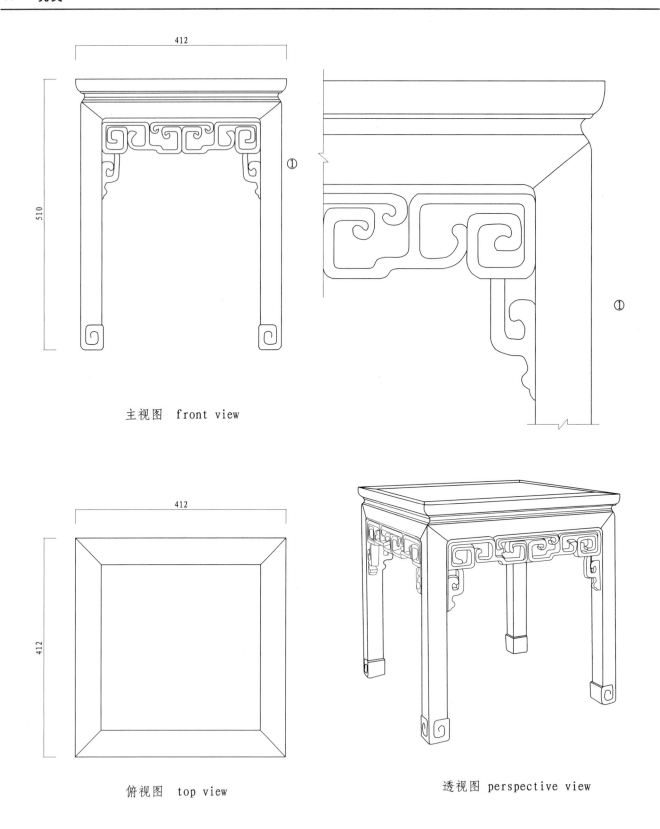

主视图 front view

俯视图 top view

透视图 perspective view

清代红木攉脚花牙方凳
Qing dynasty *hong* wood square stool with *huojiao* stretchers and flower design aprons

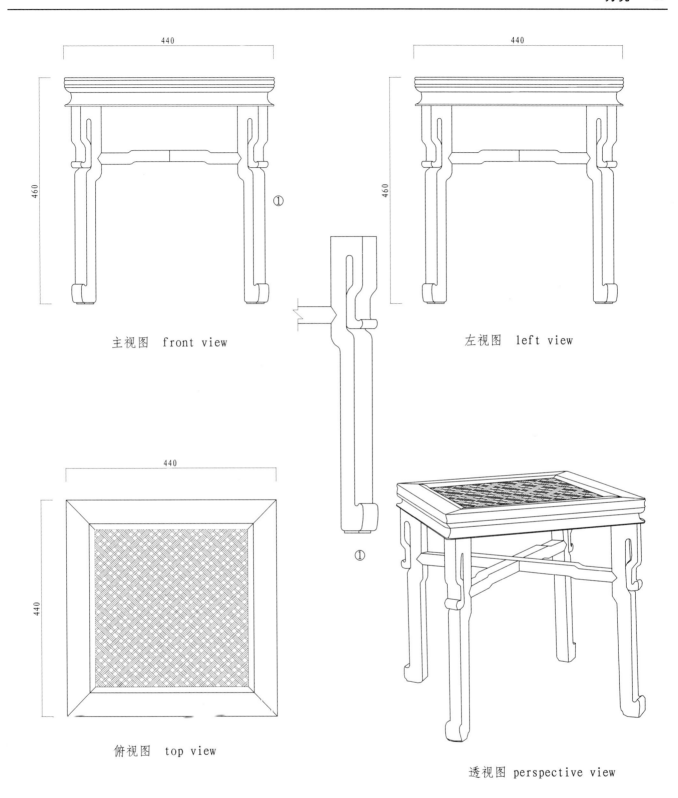

主视图 front view

左视图 left view

俯视图 top view

透视图 perspective view

清代红木藤面方凳
Qing dynasty *hong* wood square stool with mat seat

凳类

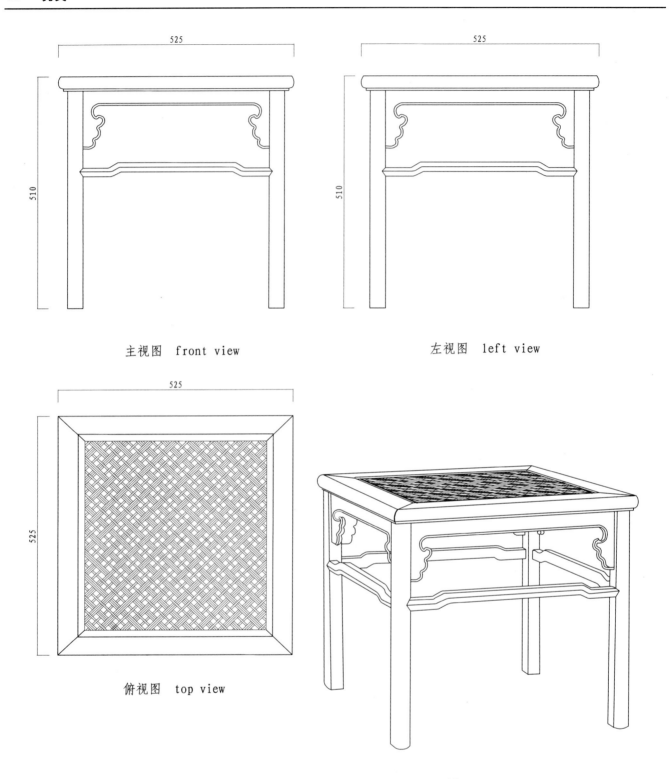

主视图 front view

左视图 left view

俯视图 top view

透视图 perspective view

明代藤面香红木方凳
Ming dynasty fragrant *hong* wood square stool with mat seat

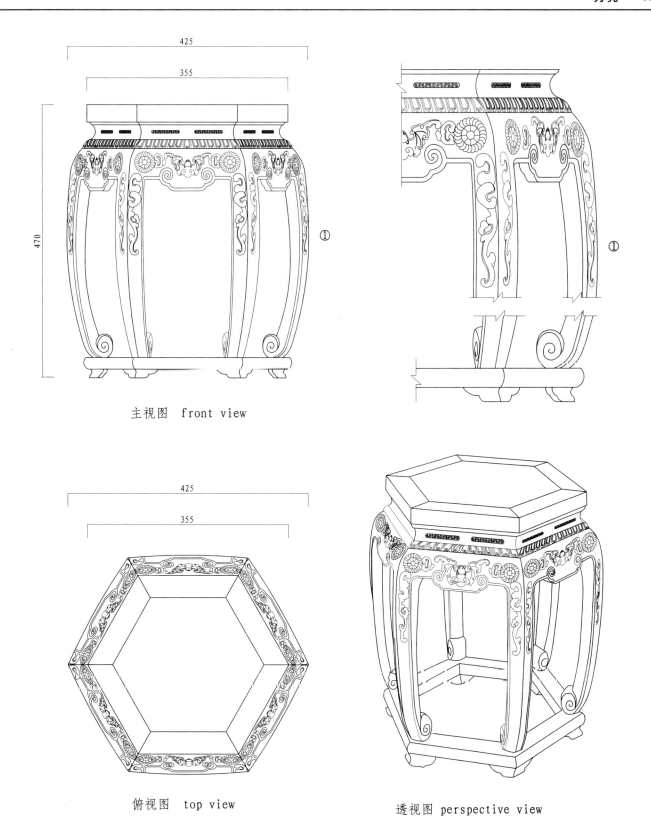

主视图 front view

俯视图 top view

透视图 perspective view

清代紫檀嵌玉团花纹六方凳
Qing dynasty *zitan* wood hexagonal stool with jade inlay and flower pattern design in medallion

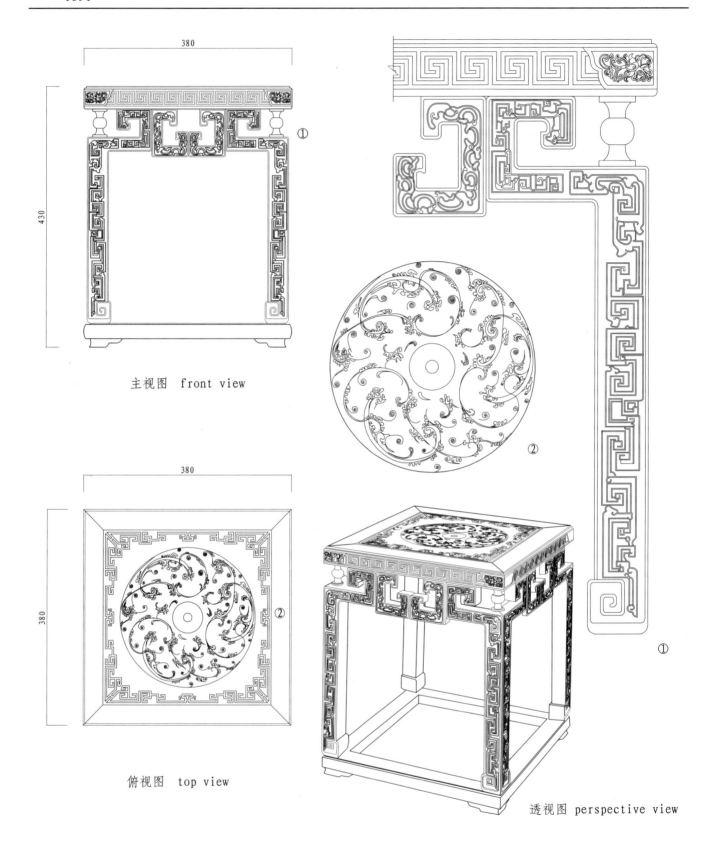

主视图 front view

俯视图 top view

透视图 perspective view

清乾隆紫檀嵌珐琅漆面团花纹方凳
Qing dynasty Qianlong period *zitan* wood lacquer square stool
with enamel inlay and flower pattern design in medallion

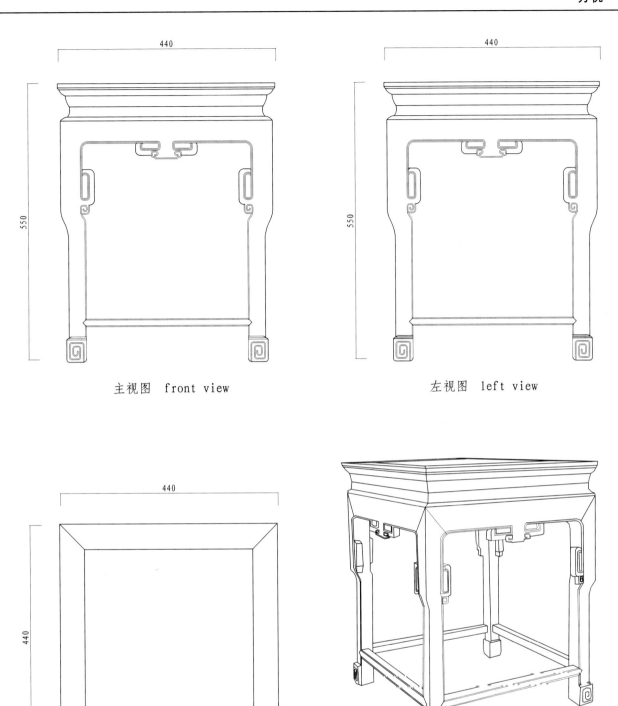

主视图 front view

左视图 left view

俯视图 top view

透视图 perspective view

明代花梨木桥梁档方杌子
Ming dynasty *huali* wood square stool with bridged stretchers

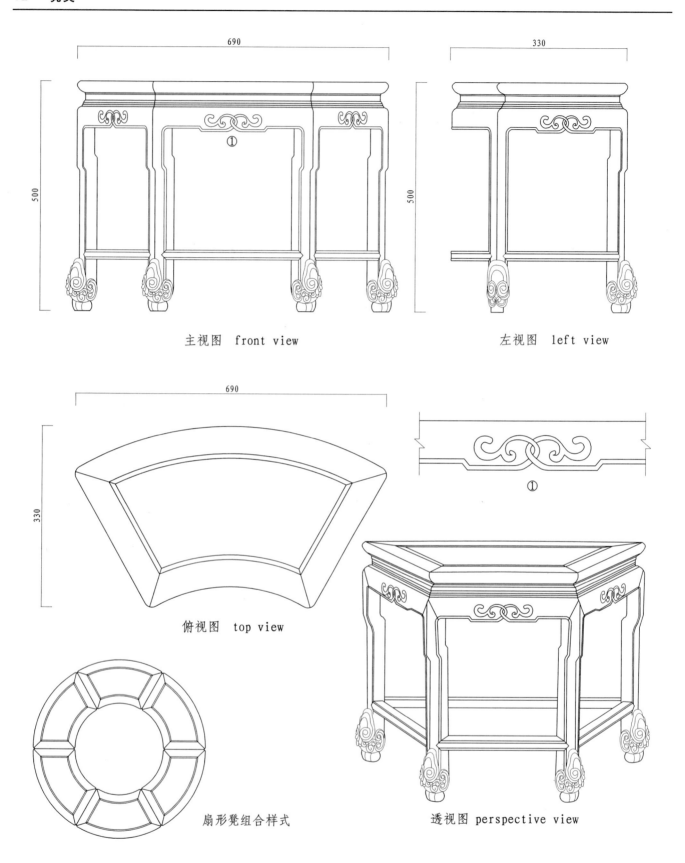

主视图 front view

左视图 left view

俯视图 top view

扇形凳组合样式

透视图 perspective view

清中期红木扇面形凳
Mid Qing dynasty *hong* wood fan-shaped stool

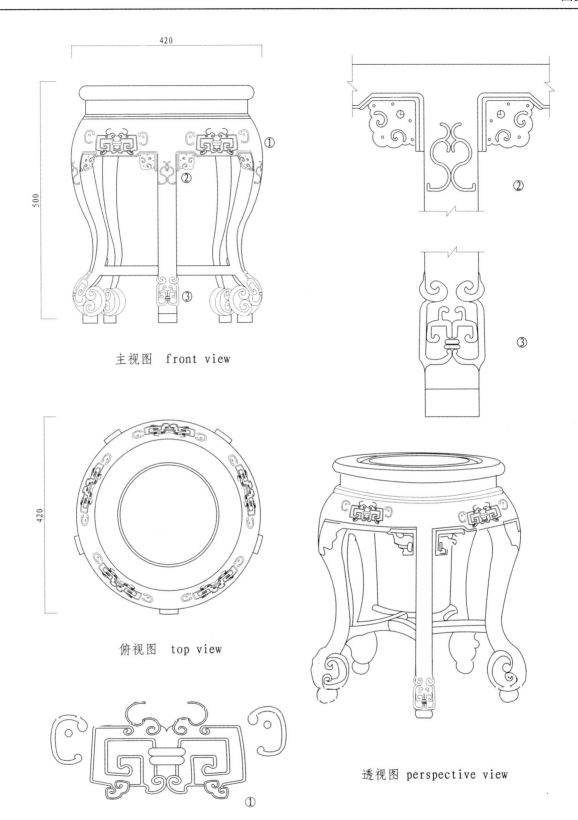

主视图 front view

俯视图 top view

透视图 perspective view

清代红木圆凳
Qing dynasty *hong* wood round stool

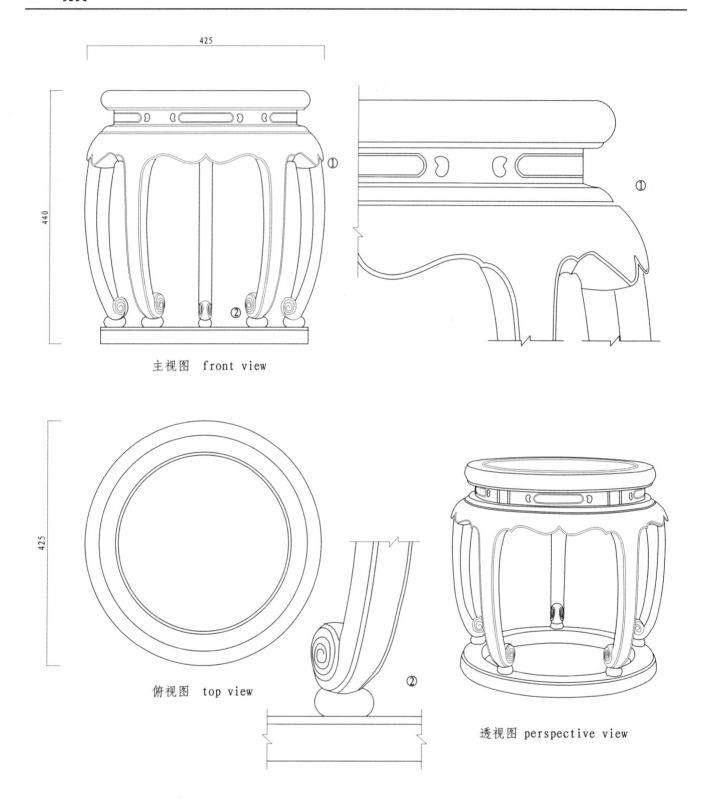

明代红漆嵌珐琅面龙戏珠纹圆凳
Ming dynasty red lacquer round stool with enamel inlay on the seat and dragons palying a pearl motif

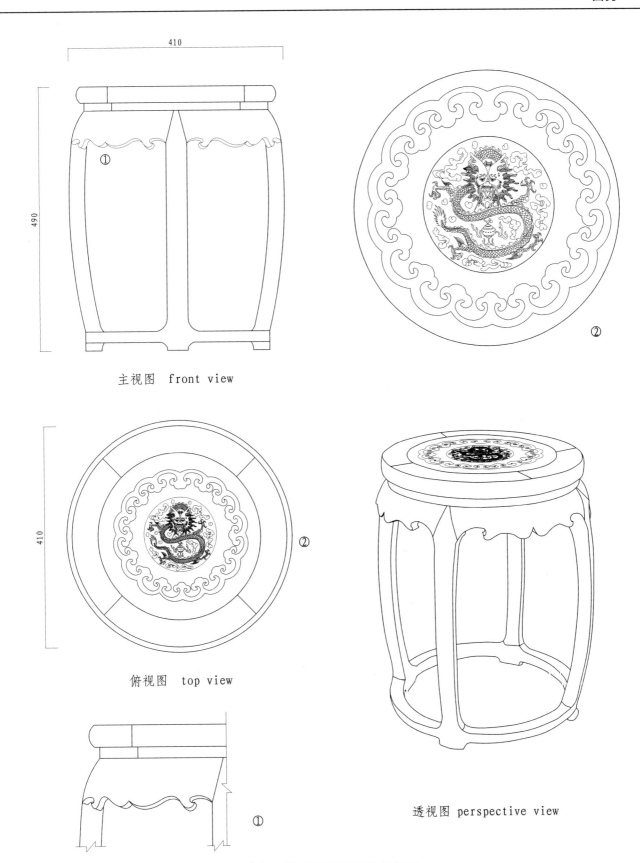

主视图 front view

俯视图 top view

透视图 perspective view

清康熙楠木嵌瓷面云龙纹圆凳
Qing dynasty Kangxi period *nan* wood round stool with porcelain inlay on the seat and cloud-dragon motif

主视图 front view

俯视图 top view

透视图 perspective view

清乾隆黄花梨玉璧纹圆凳
Qing dynasty Qianlong period *huanghuali* wood round stool
with round flat pieces of jade design

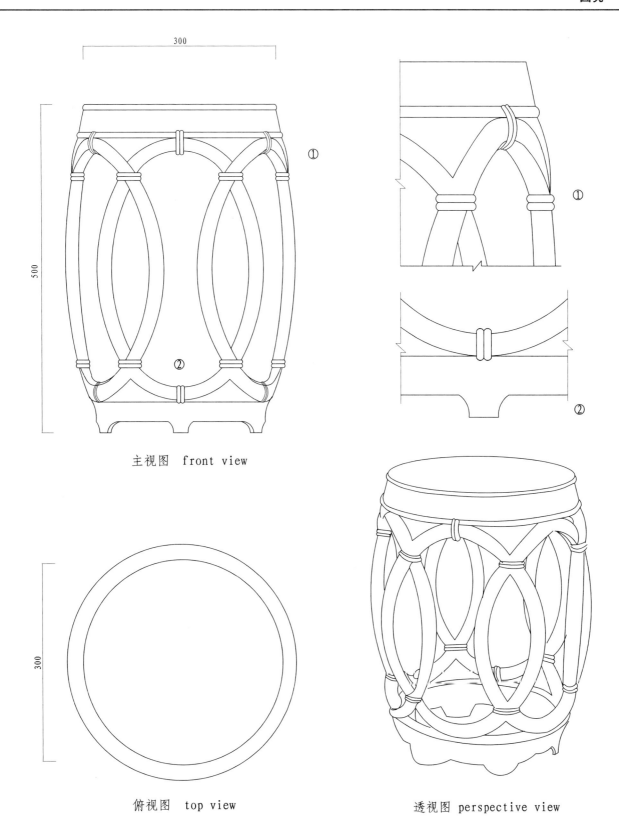

清代酸枝双钱纹鼓凳
Qing dynasty *suanzhi* wood drum stool with double-coin motif

主视图 front view

左视图 left view

俯视图 top view

透视图 perspective view

清代酸枝木卷草纹方桌
Qing dynasty *suanzhi* wood square table with curling tendril pattern

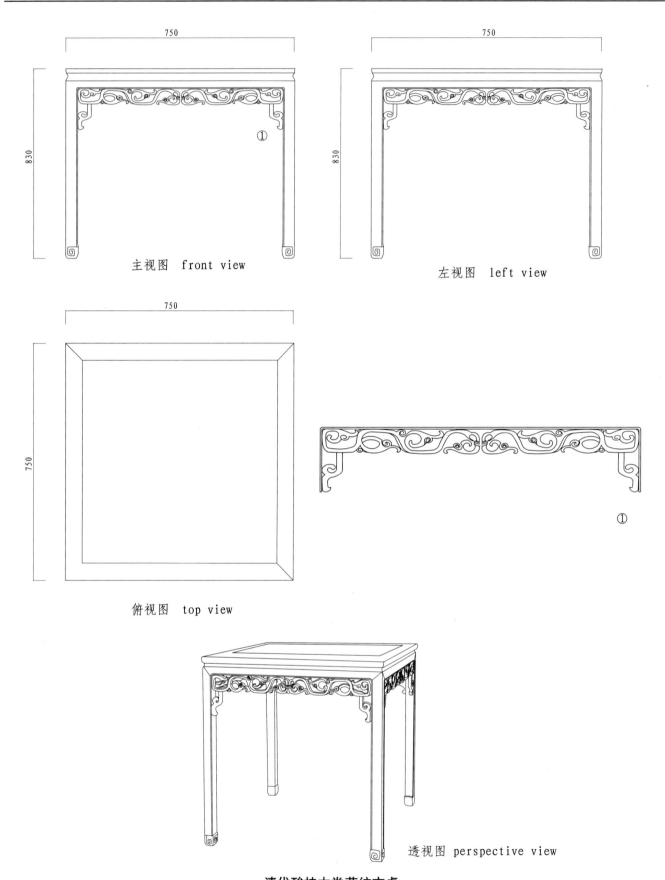

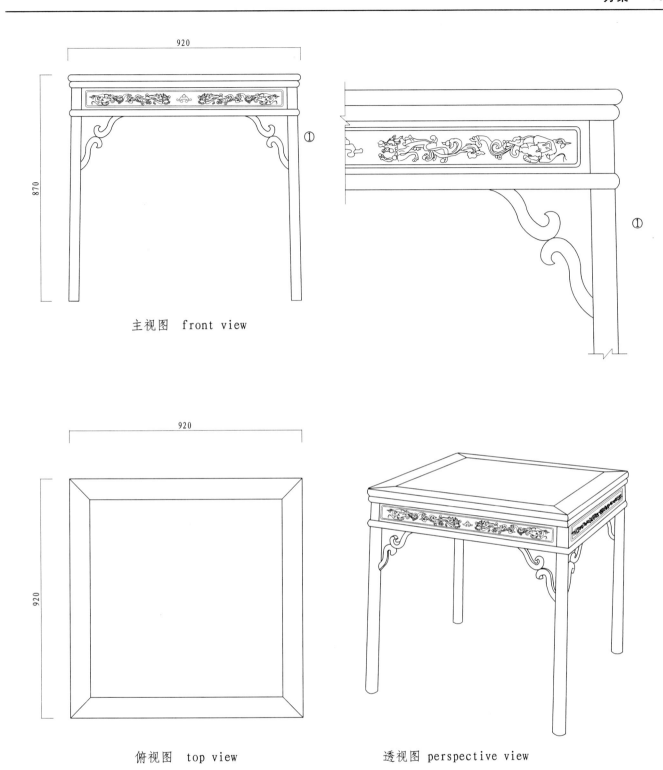

主视图 front view

俯视图 top view

透视图 perspective view

明代黄花梨螭纹方桌
Ming dynasty *huanghuali* wood square table with stylized hornless dragon design

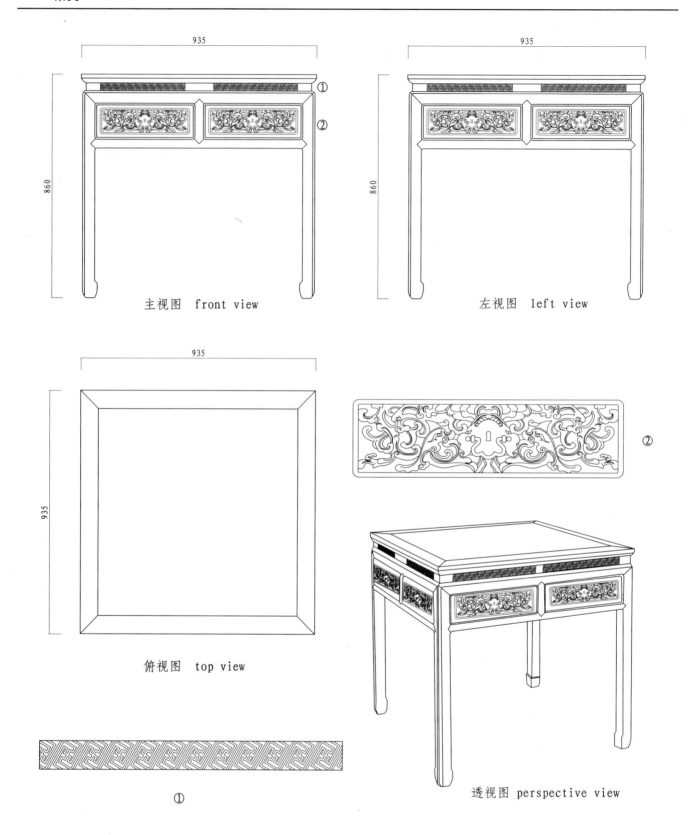

清代花梨木抽屉云石面方桌
Qing dynasty *huali* wood square table with drawers and cloud pattern marble top

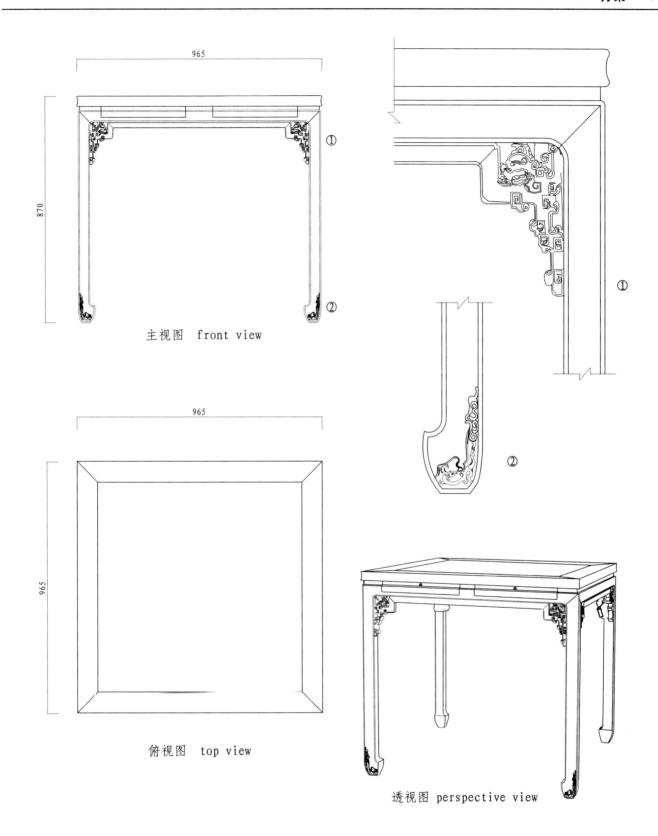

主视图 front view

俯视图 top view

透视图 perspective view

清早期紫檀夔纹暗屉方桌
Early Qing dynasty *zitan* wood square table with *kui*-dragon motif and hiding drawers

主视图 front view

左视图 left view

俯视图 top view

透视图 perspective view

清代榉木黑漆方桌
Qing dynasty *ju* wood square table with black lacquer

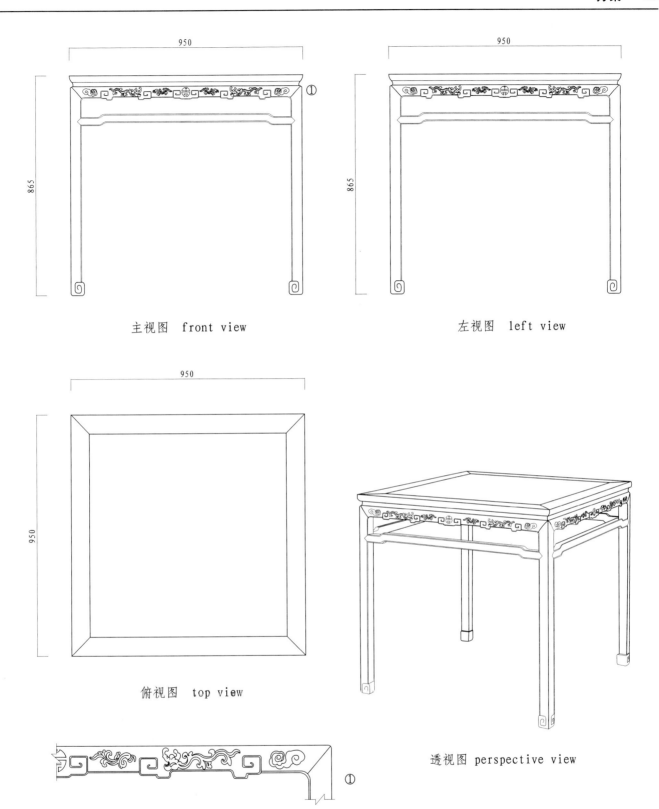

主视图 front view

左视图 left view

俯视图 top view

透视图 perspective view

清代黄花梨云龙寿字纹方桌
Qing dynasty *huanghuali* wood square table with cloud-dragon and *shou* character motif

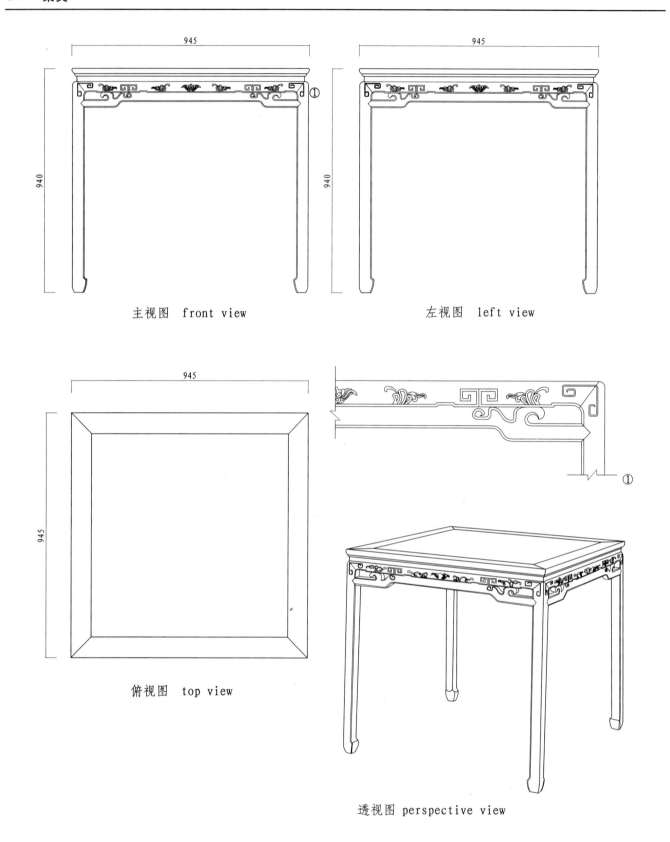

明代黄花梨卷草纹方桌
Ming dynasty *huanghuali* wood square table with curling tendril design

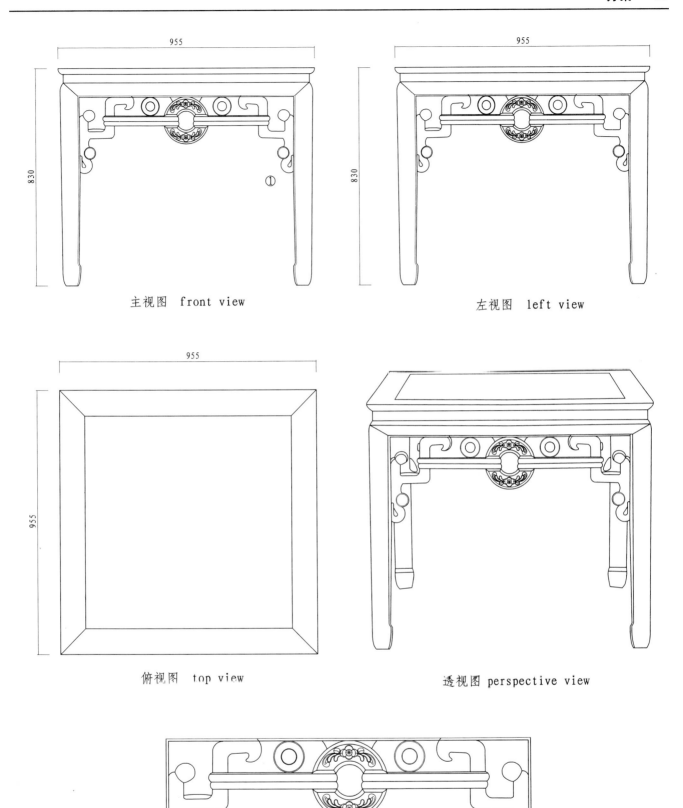

清代拱璧款束腰八仙桌
Qing dynasty *gong-bi* style waisted Eight Immortals table

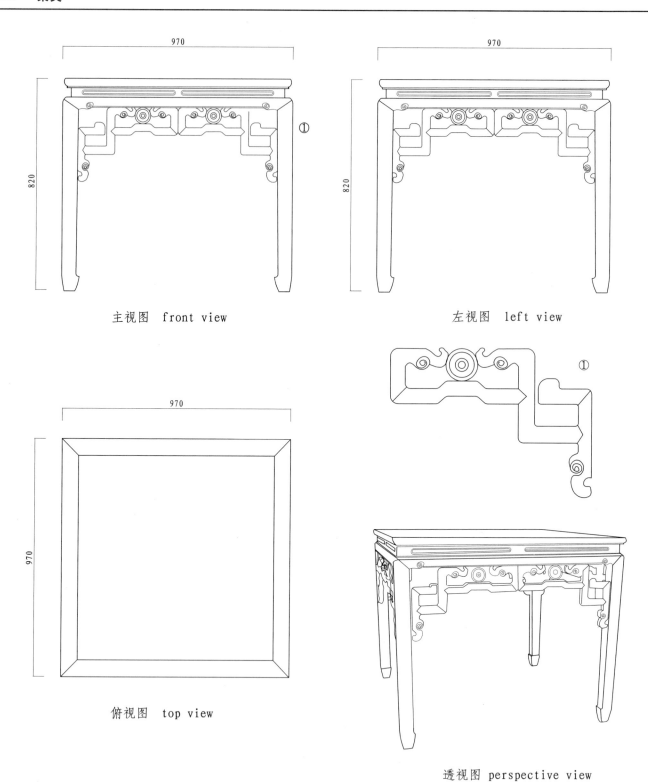

主视图 front view

左视图 left view

俯视图 top view

透视图 perspective view

清代红木八仙桌
Qing dynasty *hong* wood Eight Immortals table

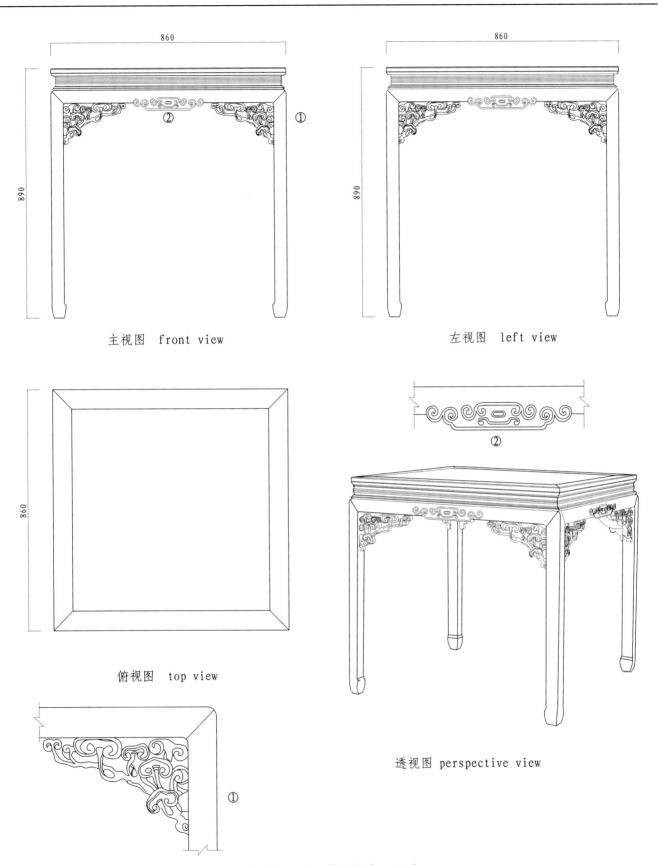

清中期红木灵芝纹插角八仙桌
Mid Qing dynasty *hong* wood Eight Immortals table with *lingzhi* fungus motif and inserted shoulder joint

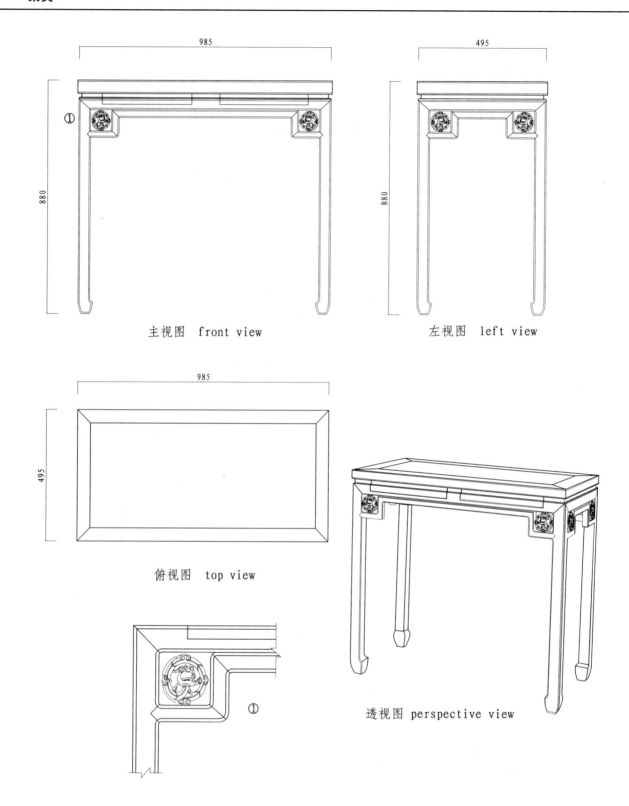

主视图 front view

左视图 left view

俯视图 top view

透视图 perspective view

明代紫檀团螭纹两屉长桌
Ming dynasty *zitan* wood rectangular table with two drawers and stylized hornless dragon design in medallion

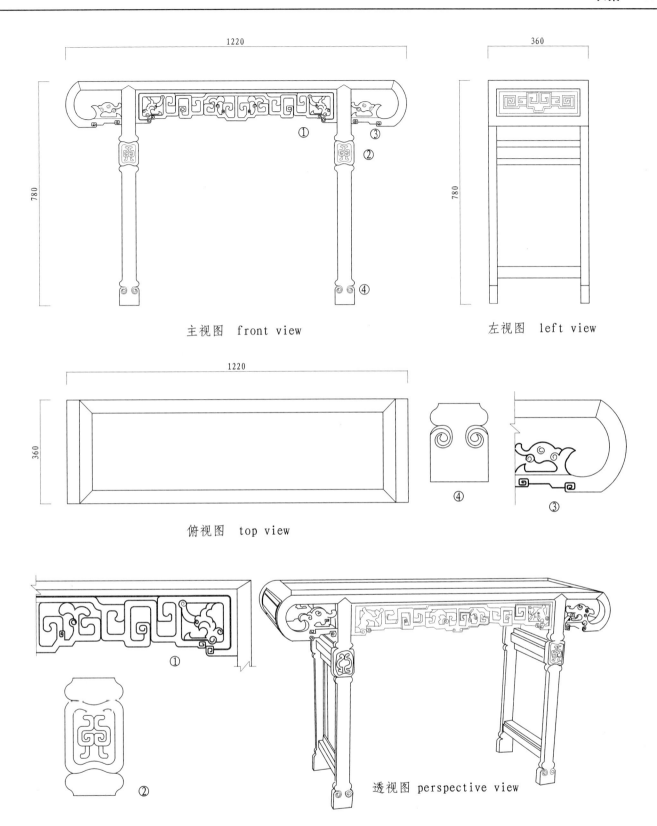

清代酸枝书卷式夔龙纹雕花长桌
Qing dynasty *suanzhi* wood narrow and rectangular table with *kui*-dragon design carving and cylindrical scroll

90　桌类

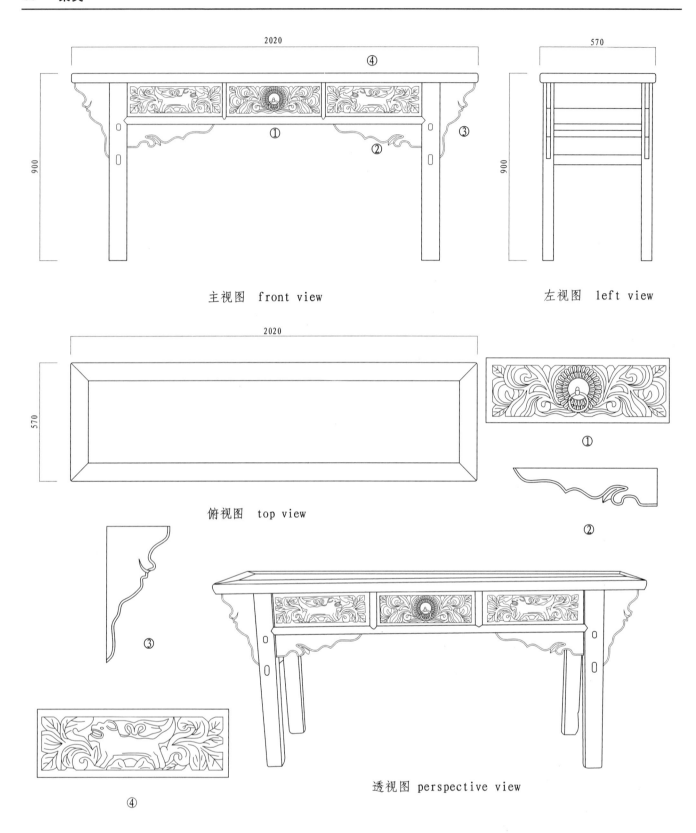

主视图　front view

左视图　left view

俯视图　top view

透视图　perspective view

明代榆木雕花三屉长桌
Ming dynasty elm wood narrow and rectangular table with three drawers and carving design

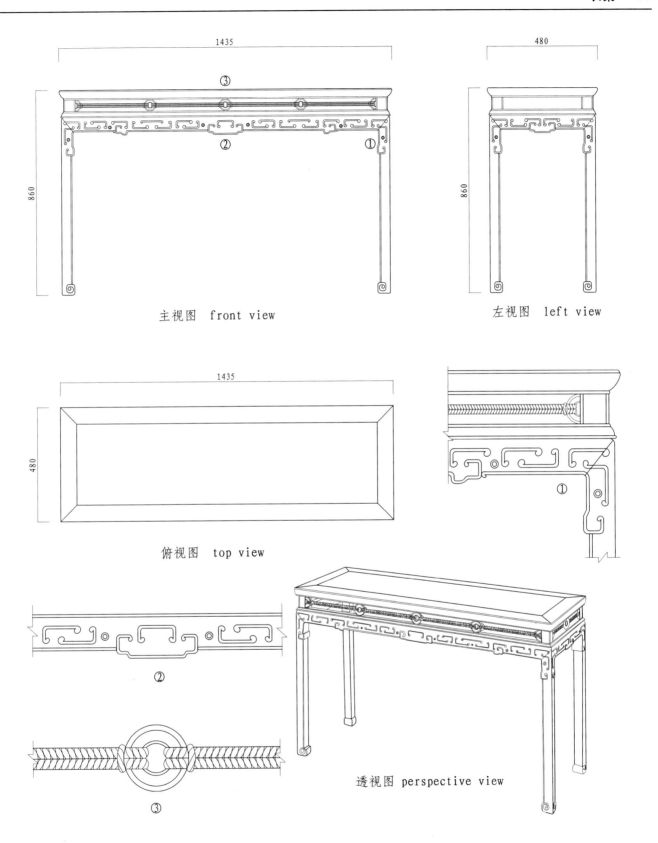

主视图 front view

左视图 left view

俯视图 top view

透视图 perspective view

清乾隆紫檀长桌
Qing dynasty Qianlong period *zitan* wood waisted narrow and rectangular table

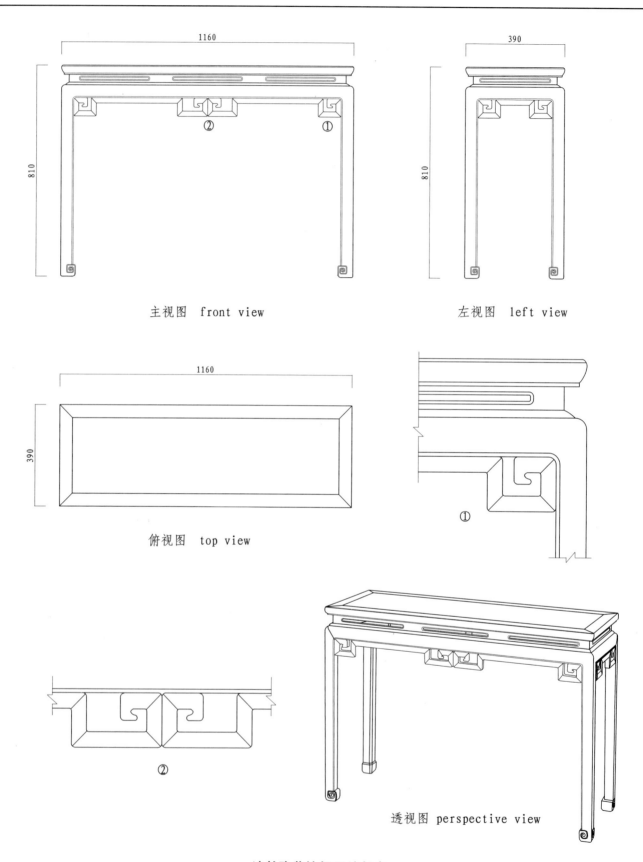

主视图 front view
左视图 left view
俯视图 top view
透视图 perspective view

清乾隆紫檀拐子纹长桌
Qing dynasty Qianlong period *zitan* wood waisted narrow and rectangular table with rectangular spiral pattern

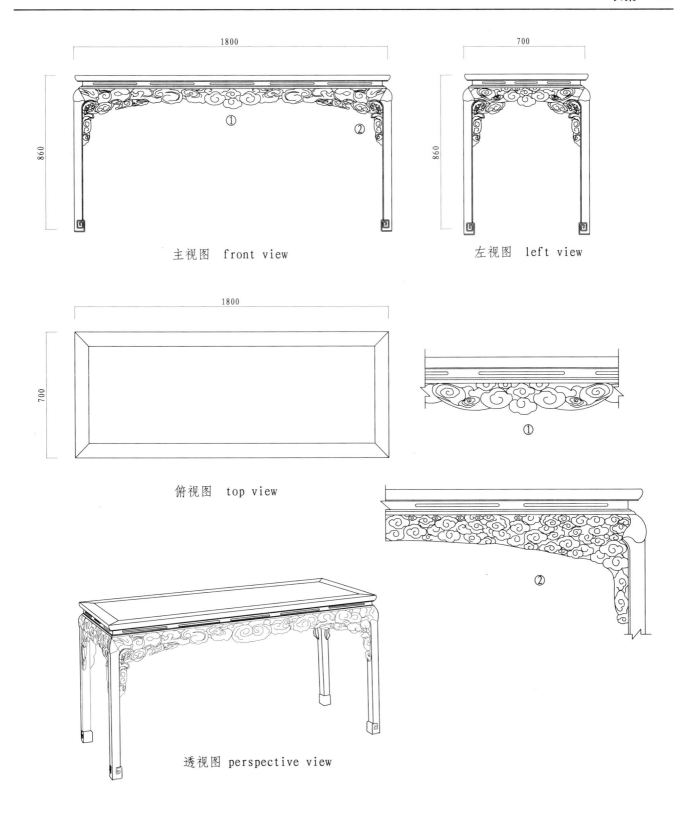

清代紫檀灵芝纹长桌
Qing dynasty *zitan* wood waisted narrow and rectangular table with *lingzhi* fungus motif

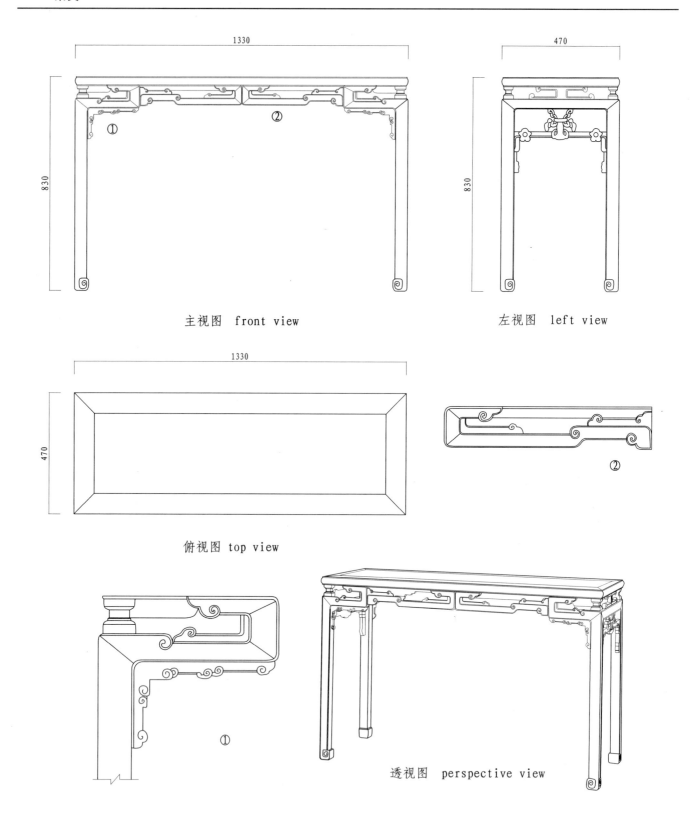

清代红木云纹长桌
Qing dynasty *hong* wood narrow and rectangular table with cloud design

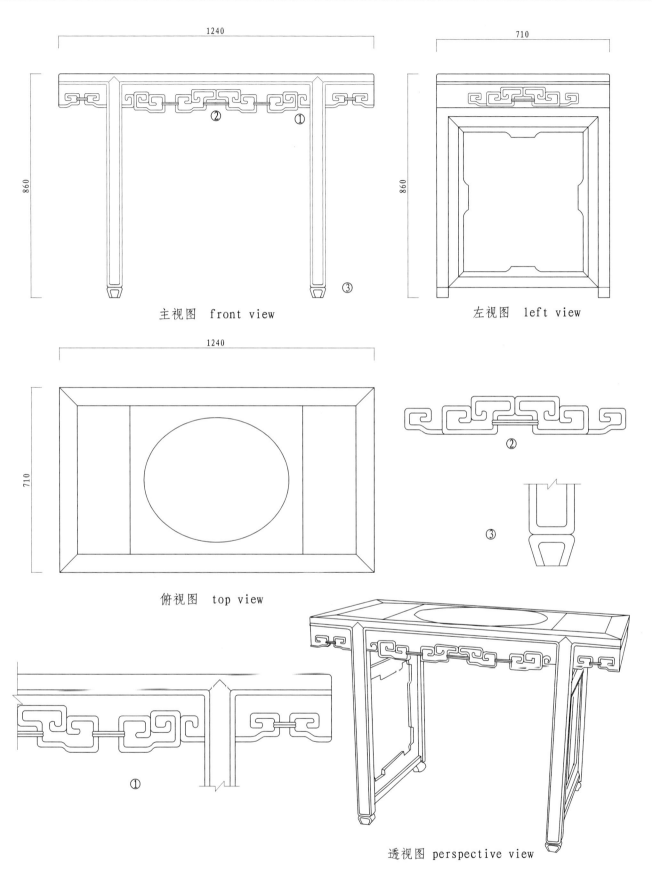

主视图 front view

左视图 left view

俯视图 top view

透视图 perspective view

清代酸枝嵌大理石长方桌
Qing dynasty *suanzhi* wood narrow and rectangular table with marble inlay

96　桌类

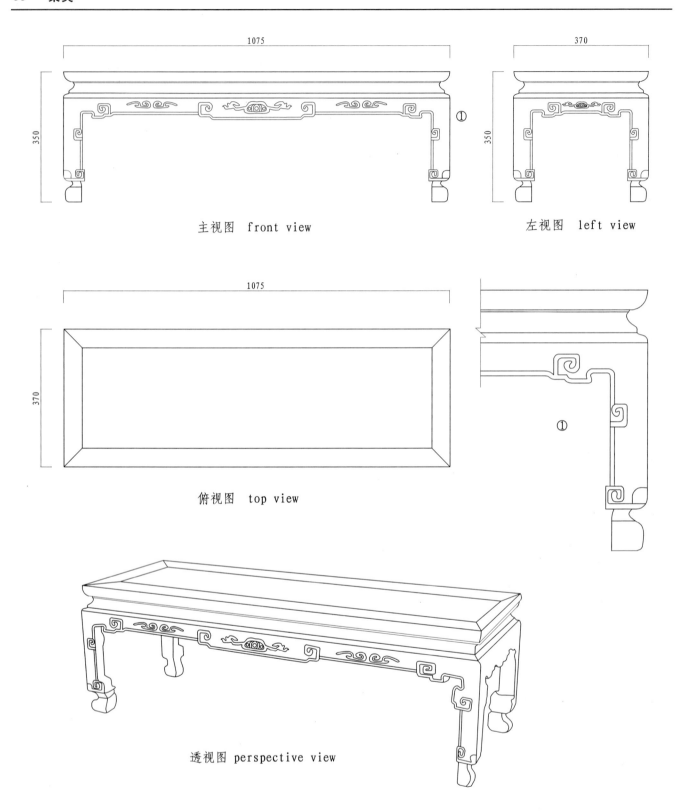

主视图　front view

左视图　left view

俯视图　top view

透视图　perspective view

清代紫檀浮雕长方桌
Qing dynasty *zitan* wood narrow and rectangular table with relief carving

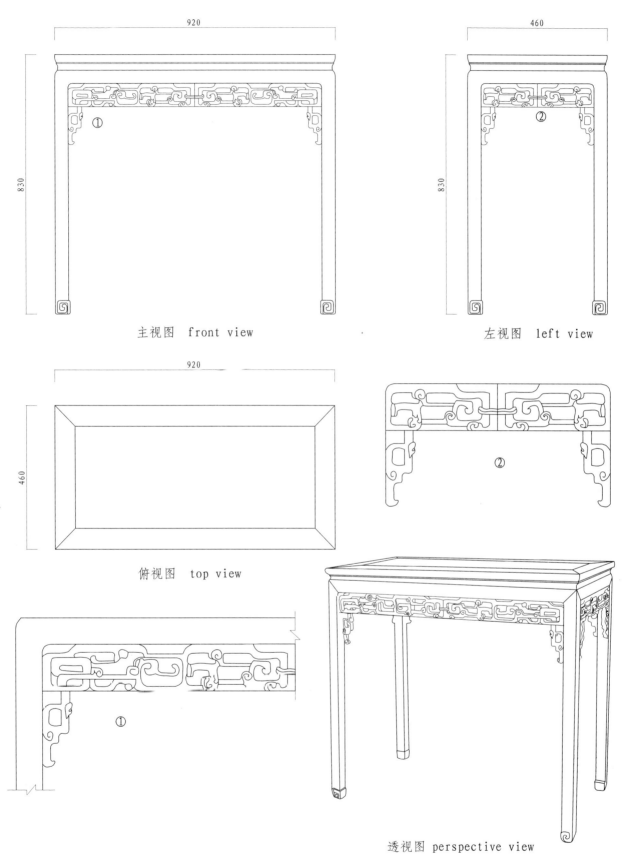

主视图 front view

左视图 left view

俯视图 top view

透视图 perspective view

清代红木半桌
Qing dynasty *hong* wood half table

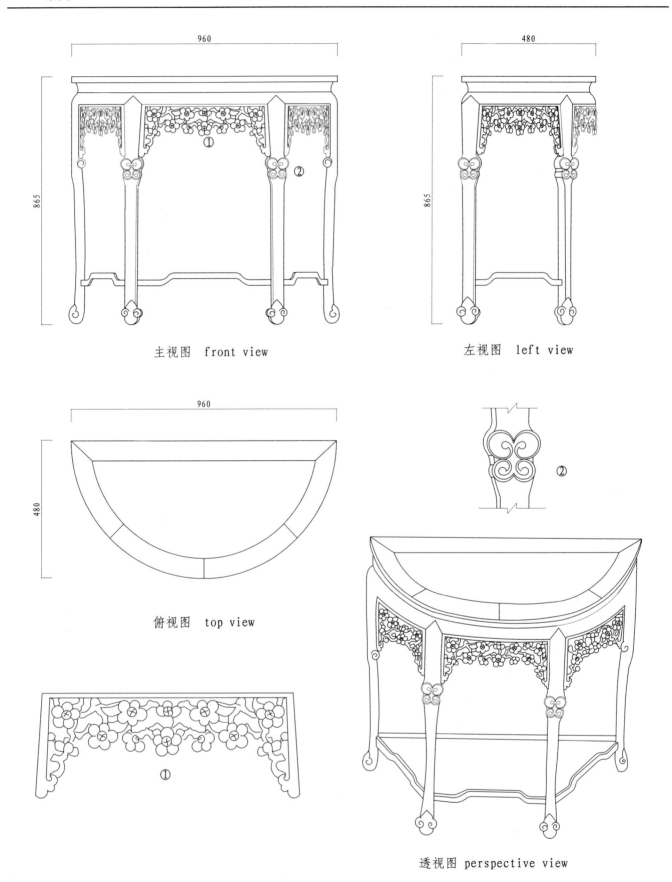

主视图 front view
左视图 left view
俯视图 top view
透视图 perspective view

清代紫檀木圆形半桌
Qing dynasty *zitan* wood half round table

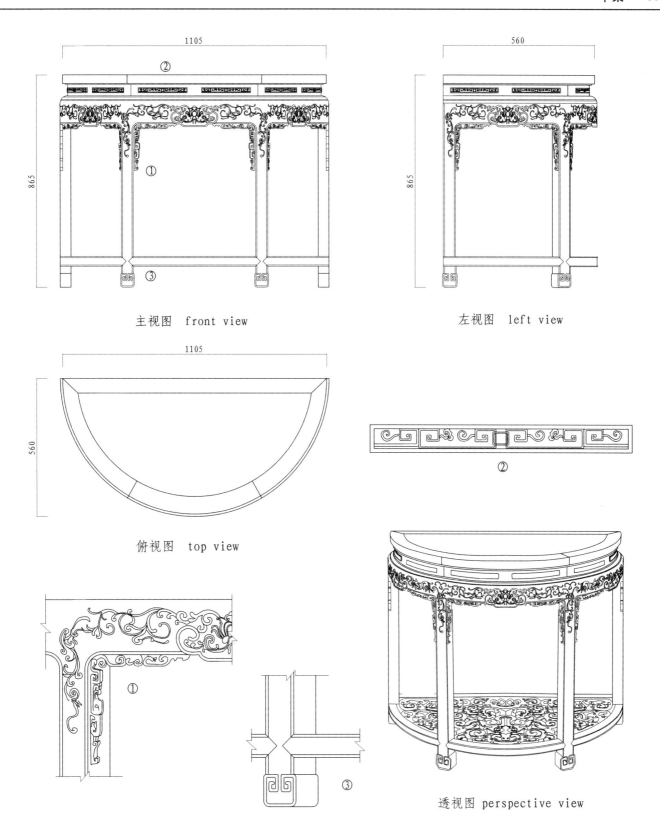

主视图 front view

俯视图 top view

左视图 left view

透视图 perspective view

清代紫檀番莲纹半圆桌
Qing dynasty *zitan* wood half round table with passion flower pattern

主视图 front view　左视图 left view

俯视图 top view

透视图 perspective view

清代禧寿福纹红木半桌
Qing dynasty hond wood half table with xi-shou-fu character design

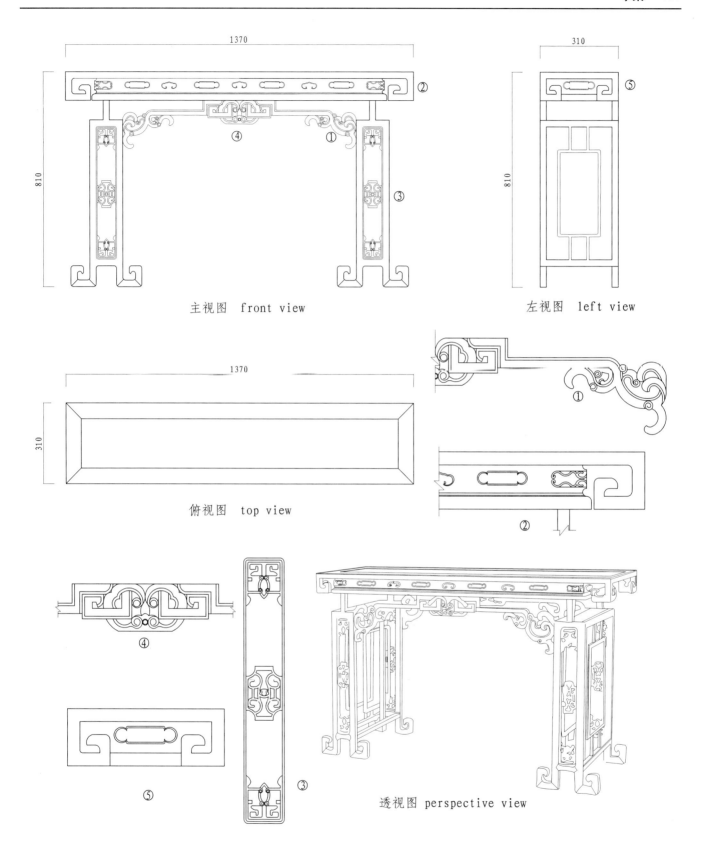

主视图 front view

左视图 left view

俯视图 top view

透视图 perspective view

清中期红木琴桌
Mid Qing dynasty *hong* wood lute table

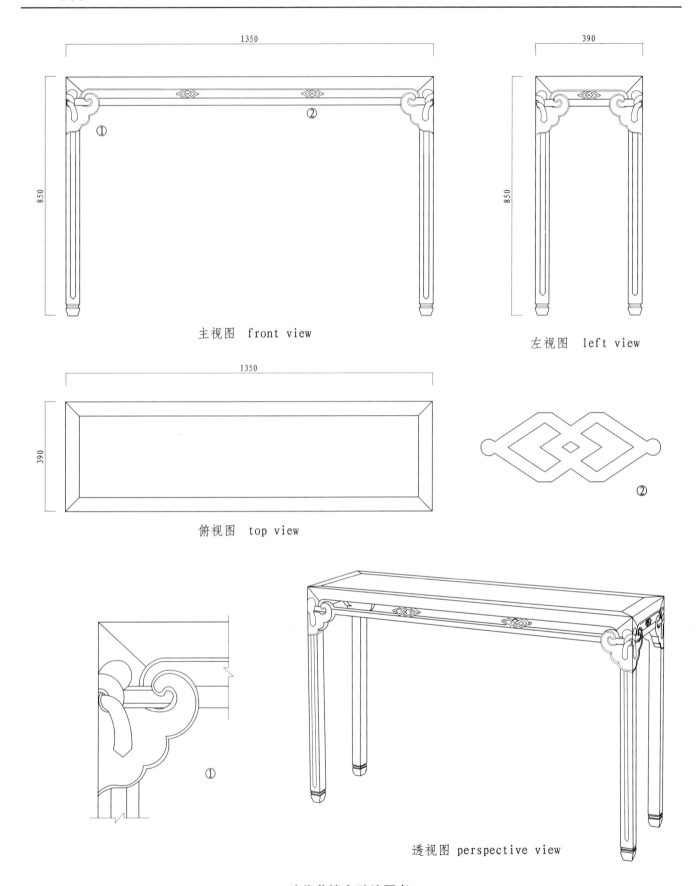

清代紫檀方胜纹琴桌

Qing dynasty *zitan* wood lute table with interlocked diamond design

主视图 front view

左视图 left view

俯视图 top view

透视图 perspective view

明代榉木四屉书桌
Ming dynasty *ju* wood recessed-leg table with four drawers

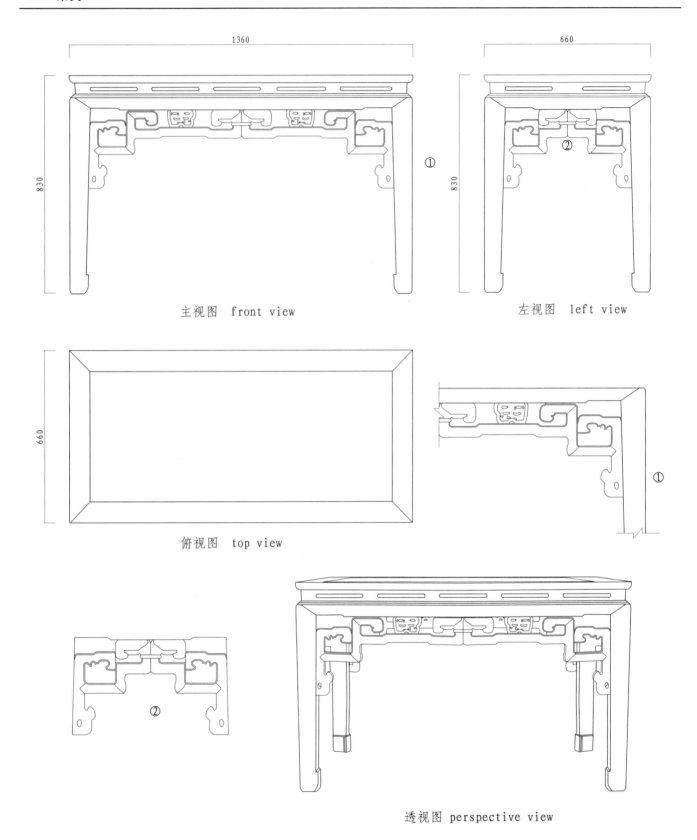

主视图 front view

左视图 left view

俯视图 top view

透视图 perspective view

清晚期花梨木拐子纹供桌
Late Qing dynasty *huali* wood altar table with rectangular spiral pattern

主视图 front view
左视图 left view
俯视图 top view
透视图 perspective view

明末清初黄花梨带底座供桌
Late Ming or Early Qing dynasty *huanghuali* wood altar table with a base

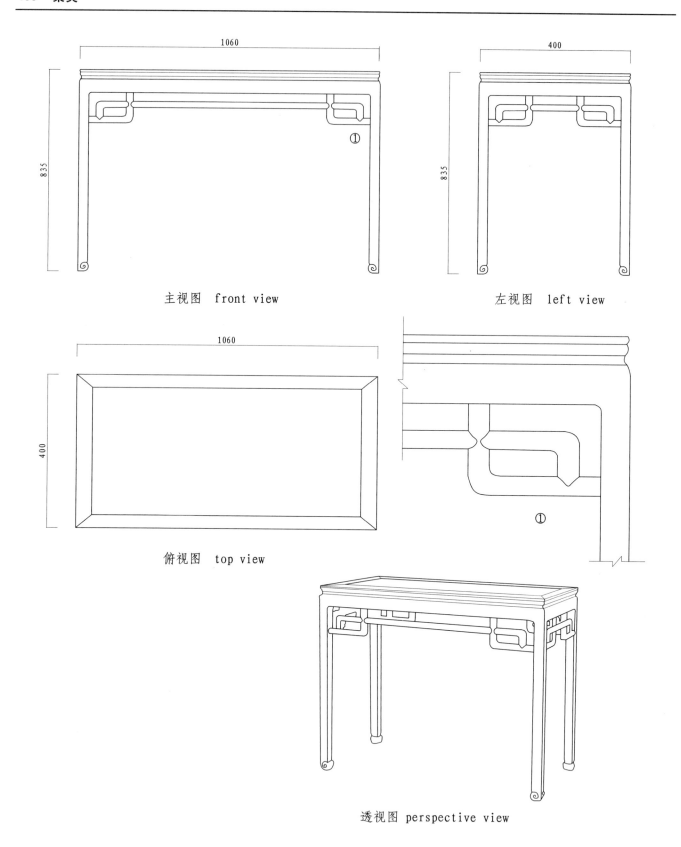

主视图 front view

左视图 left view

俯视图 top view

透视图 perspective view

明代黄花梨条桌
Ming dynasty *huanghuali* wood narrow rectangular table with corner legs

条桌 107

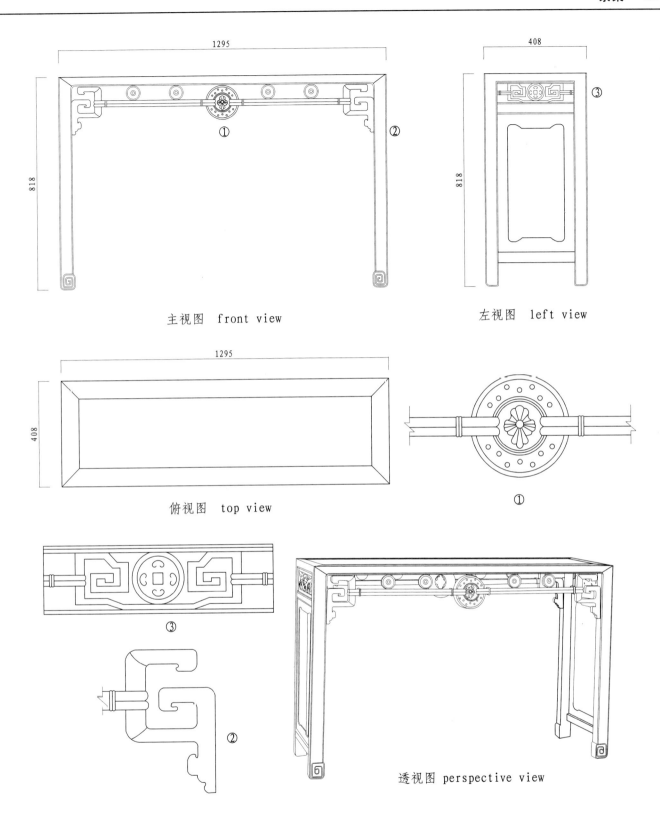

清代红木绳璧档条桌
Qing dynasty *hong* wood narrow rectangular table
with corner legs and rope-jade design stretchers

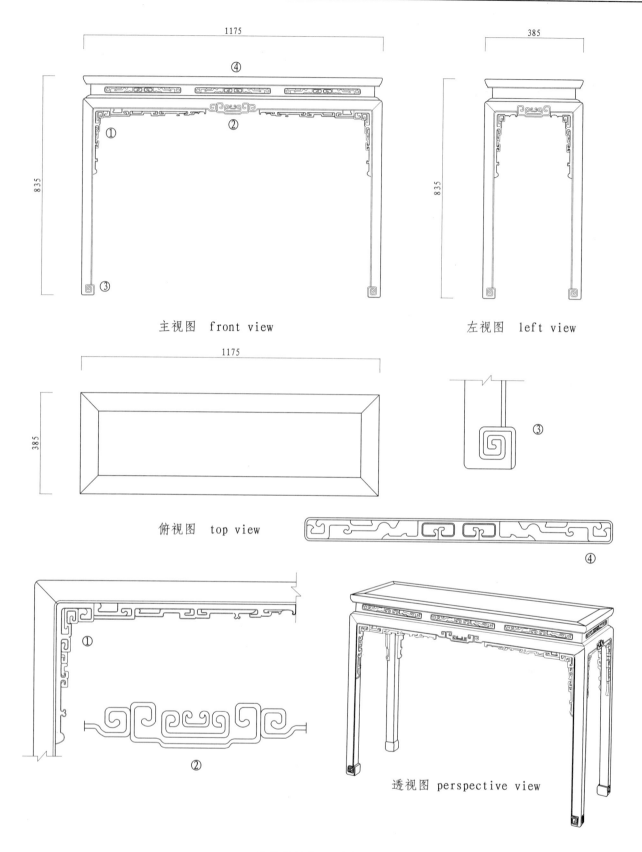

主视图 front view　　左视图 left view

俯视图 top view

透视图 perspective view

清代紫檀拐子纹条桌
Qing dynasty *zitan* wood narrow rectangular table with corner legs and rectangular spiral pattern

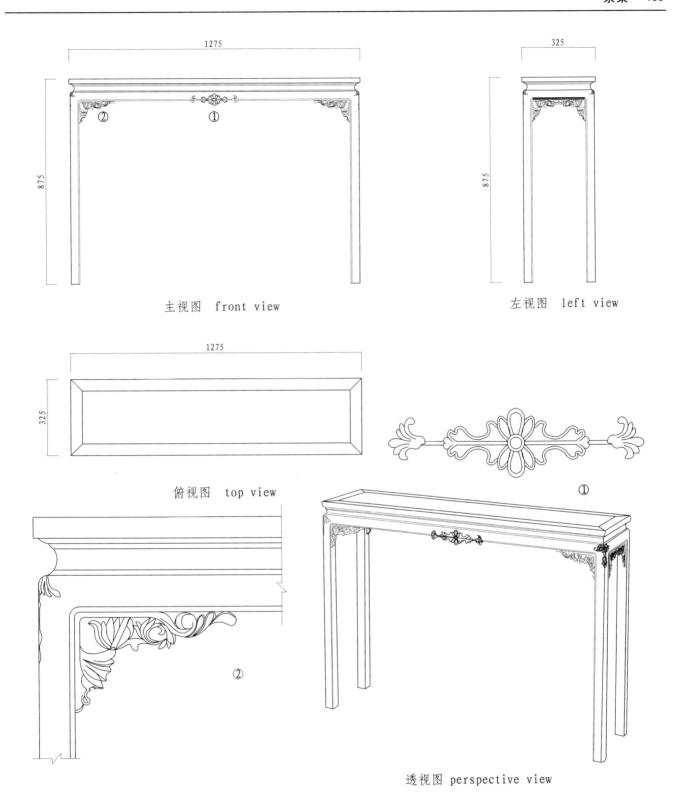

清乾隆紫檀西番莲纹条桌
Qing dynasty Qianlong period *zitan* wood narrow rectangular table with corner legs and passion flowers pattern

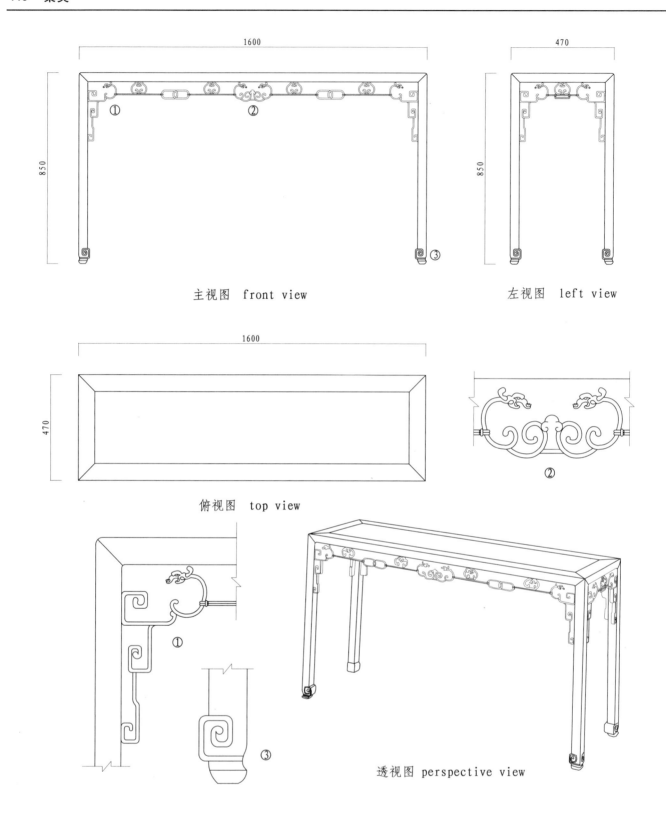

主视图 front view 左视图 left view

俯视图 top view

透视图 perspective view

清乾隆柏木玉宝珠纹条桌
Qing dynasty Qianlong period cypress wood narrow rectangular table with corner legs and jadepearls pattern

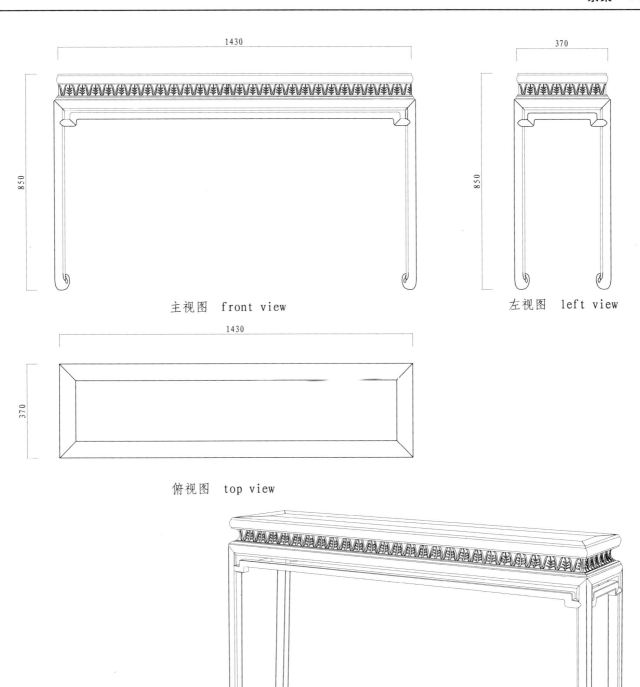

主视图 front view

左视图 left view

俯视图 top view

透视图 perspective view

清代紫檀蕉叶纹条桌
Qing dynasty *zitan* wood narrow rectangular table with corner legs and banana leaves pattern

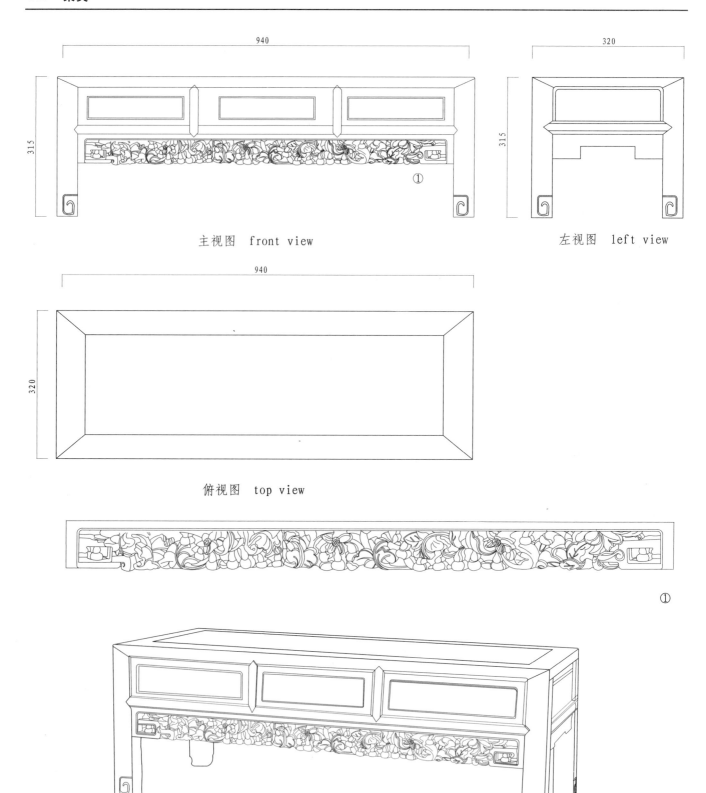

主视图 front view

左视图 left view

俯视图 top view

透视图 perspective view

清代酸枝木葫芦纹三屉炕琴桌
Qing dynasty *suanzhi* wood *kang* lute table with three drawers and gourd motif

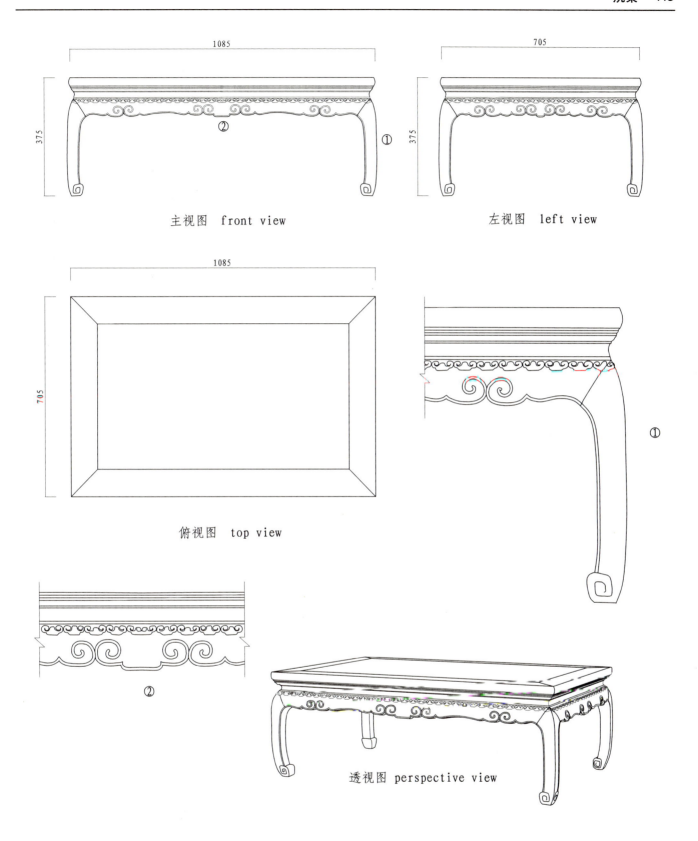

清早期紫檀卷云纹炕桌
Early Qing dynasty *ziatn* wood *kang* table with scrolled cloud motif

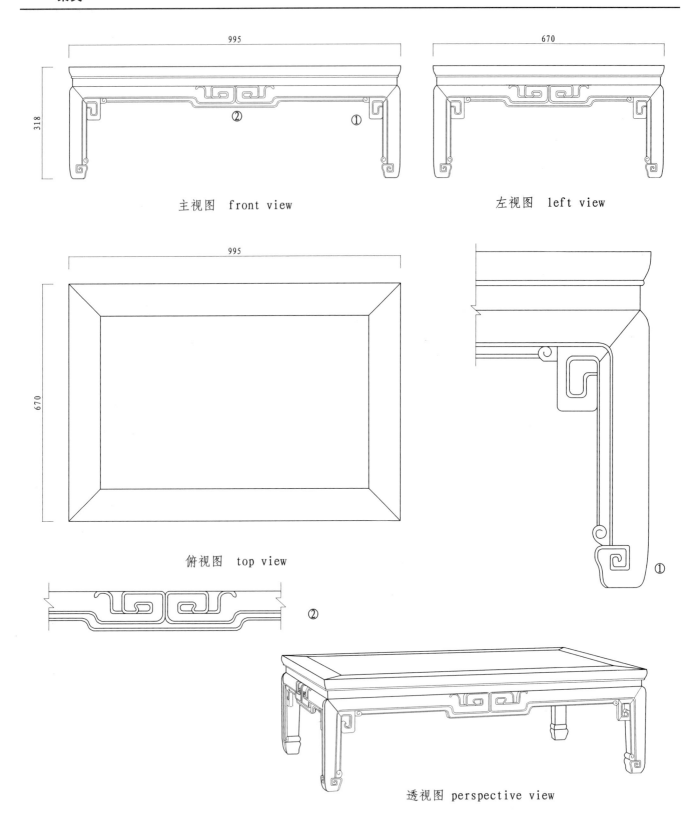

清代束腰镂空牙条炕桌
Qing dynasty waisted *kang* table with openwork carving on aprons

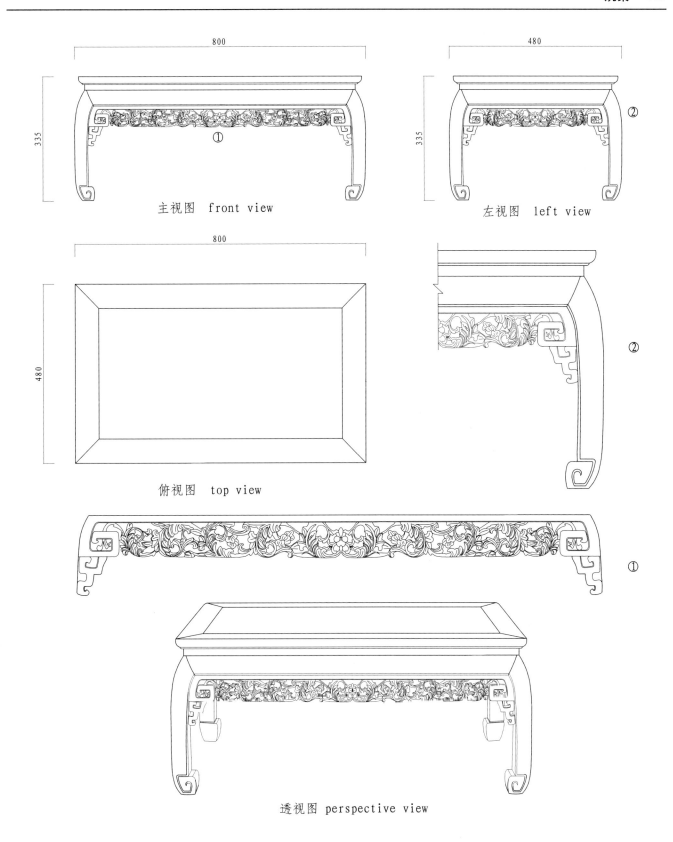

清代花梨木透雕卷草纹炕桌
Qing dynasty *huali* wood *kang* table with openwork carving of curling tendril design

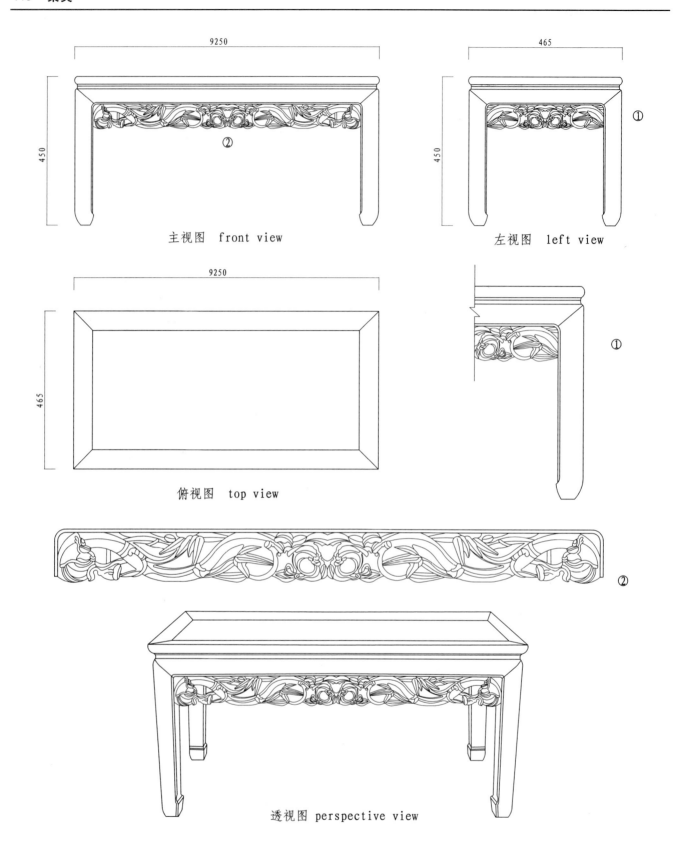

清代红木桃竹纹炕桌
Qing dynasty *hong* wood *kang* table with peach-bamboo pattern

117　房前桌

主视图 front view

左视图 left view

俯视图 top view

透视图 perspective view

清代宁式楠木房前桌
Qing dynasty Ningbo style *nan* wood table, put in front of the window in bedroom

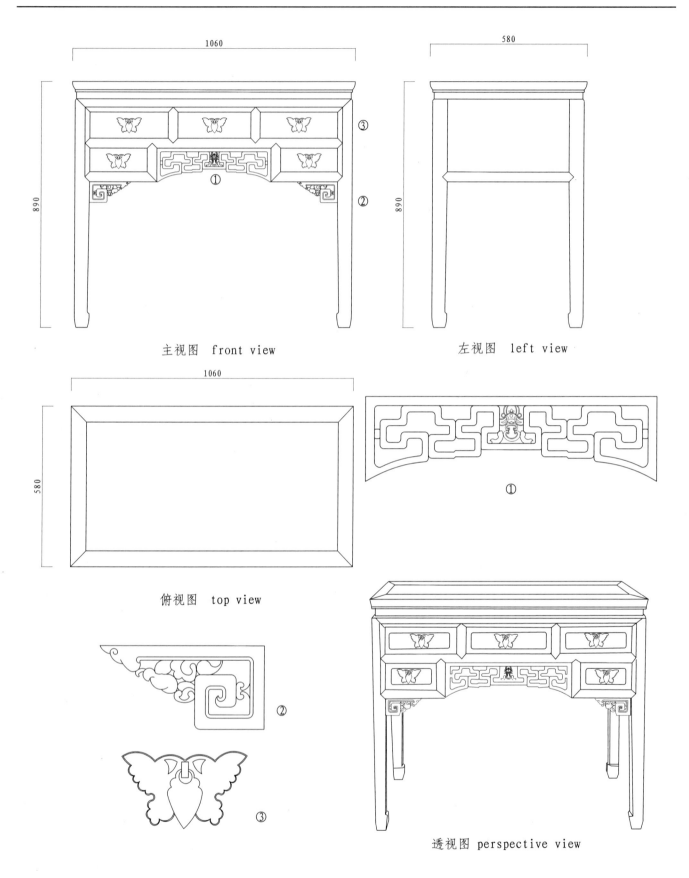

主视图 front view

左视图 left view

俯视图 top view

透视图 perspective view

清代宁式柏木房前桌
Qing dynasty Ningbo style cypress wood table, put in front of the window in bedroom

主视图 front view

左视图 left view

俯视图 top view

透视图 perspective view

清代紫檀雕花卉鱼桌
Qing dynasty *zitan* wood fish table with flowes carving design

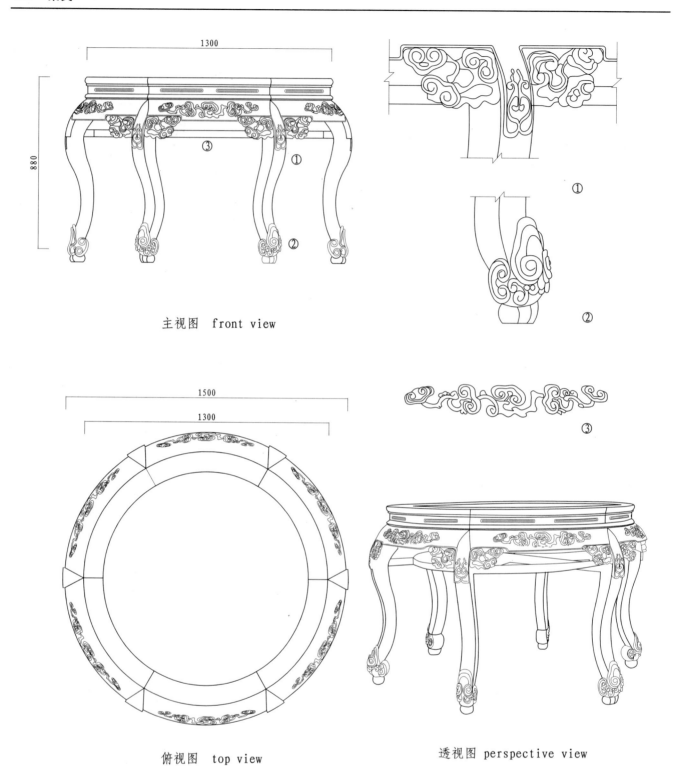

主视图　front view

俯视图　top view

透视图　perspective view

清代花梨木鼓腿膨牙大圆台
Qing dynasty *huali* wood large round table with convex aprons and bulging legs ending in horse-hoof feet

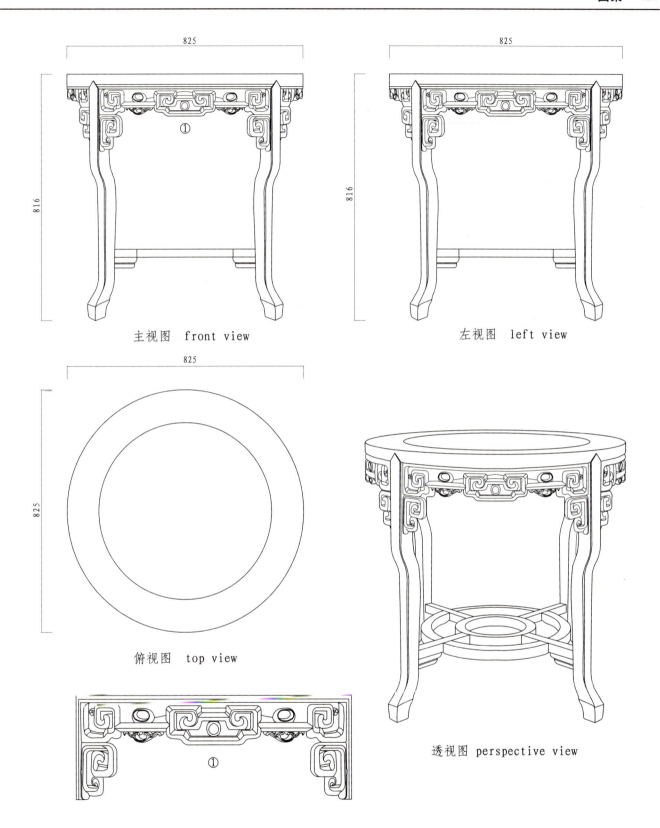

主视图 front view

左视图 left view

俯视图 top view

透视图 perspective view

清代酸枝木镶大理石拐子纹圆桌
Qing dynasty *suanzhi* wood round table with marble inlay and rectangular spiral pattern

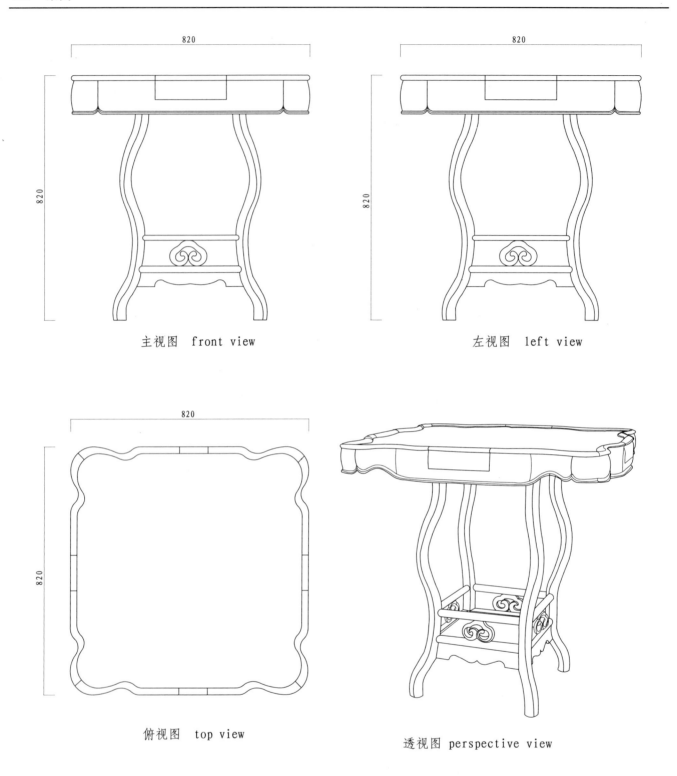

主视图 front view

左视图 left view

俯视图 top view

透视图 perspective view

清代酸枝木海棠花形麻将桌
Qing dynasty *suanzhi* wood *mah-jong* table with Chinese flowering crab-apple shaped top

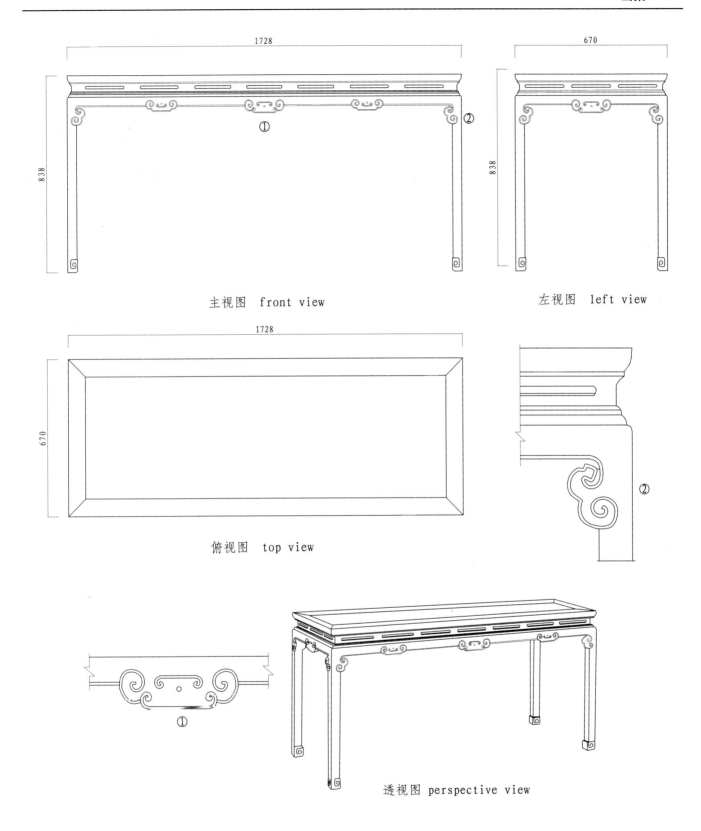

清代紫檀有束腰马蹄足画案
Qing dynasty *zitan* wood waisted recessed-leg painting table with horse-hoof feet

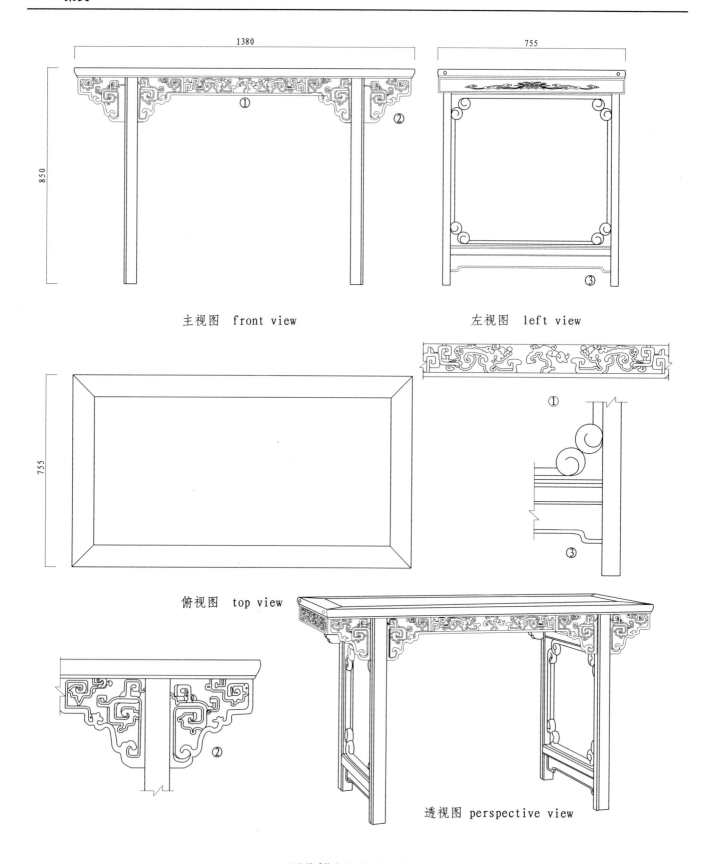

明代榉木勾卷纹画案
Ming dynasty *ju* wood recessed-leg painting table with hook-and-scroll pattern

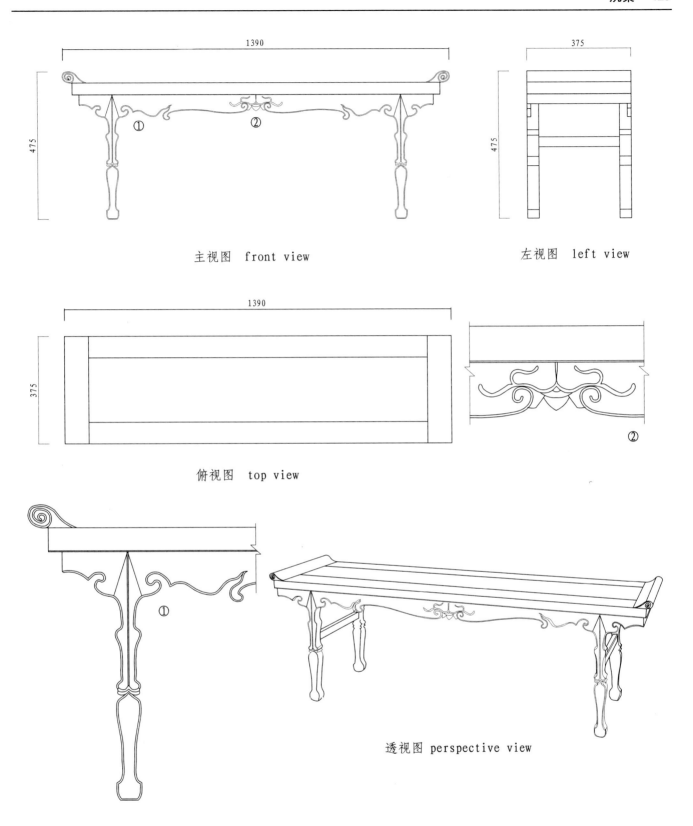

明代黄花梨翘头炕案
Ming dynasty *huanghuali* wood recessed-leg *kang* table with everted flanges

126 案类

明代花梨夔凤纹翘头案
Ming dynasty *huali* wood recessed-legs table
with everted flanges on the top and *kui*-phoenix pattern

明代黄花梨双螭纹翘头案
Ming dynasty *huanghuali* wood recessed-leg table with everted flanges and double stylized hornless dragons design

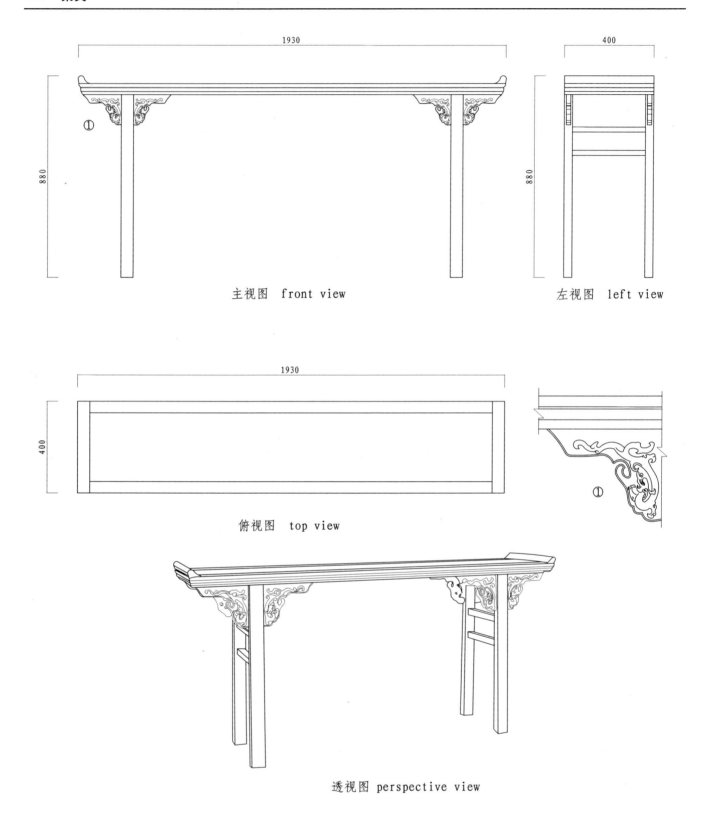

主视图 front view　　左视图 left view

俯视图 top view

透视图 perspective view

明代紫檀木翘头案
Ming dynasty *zitan* wood recessed-leg table with everted flanges

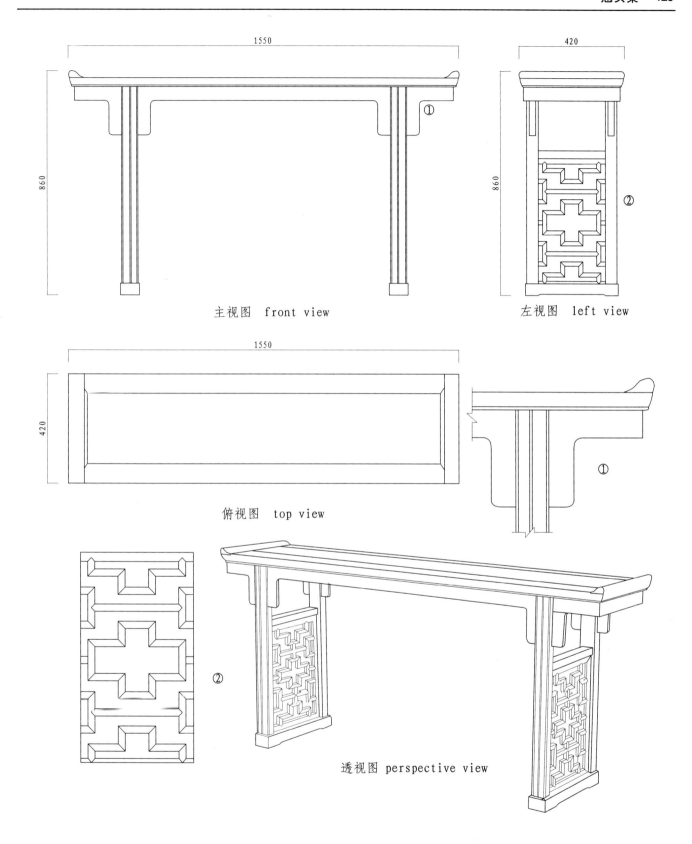

主视图 front view

左视图 left view

俯视图 top view

透视图 perspective view

明代紫檀木夹头榫带托子翘头案
Ming dynasty *zitan* wood recessed-leg table with everted flanges, elongated bridle joints and side floor stretchers

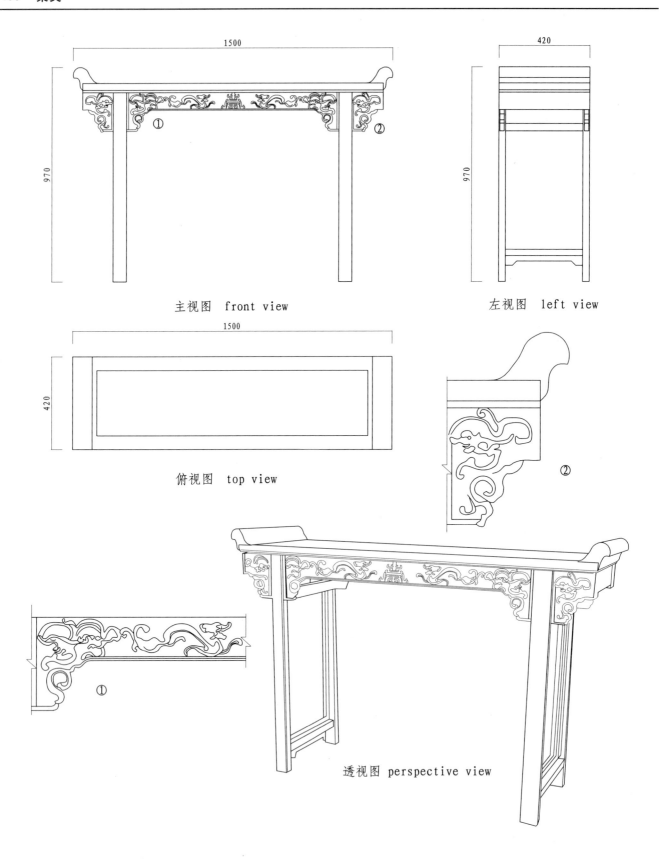

主视图 front view

左视图 left view

俯视图 top view

透视图 perspective view

明代黄花梨雕龙翘头案
Ming dynasty *huanghuali* wood recessed-leg table with everted flanges and dragon design carving

明代黄花梨雕花翘头条案
Ming dynasty *huanghuali* wood recessed-legs table with everted fanges and flowers pattern carving

清代老花梨木龙凤纹翘头案
Qing dynasty old *huali* wood recessed-leg table with everted flanges and dragon-and-phoenix design

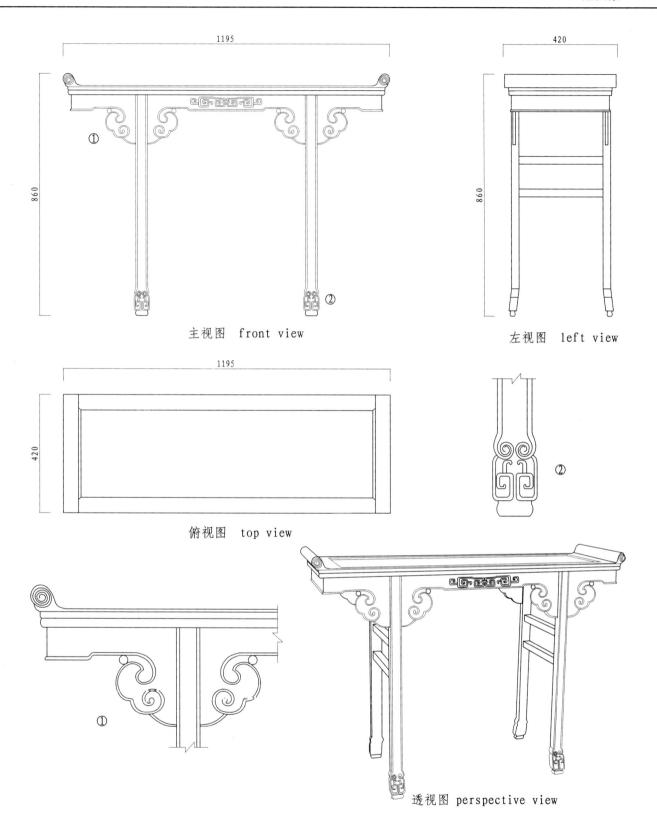

清代苏式红木小翘头案
Qing dynasty Suzhou style *hong* wood small recessed-leg table with everted flanges

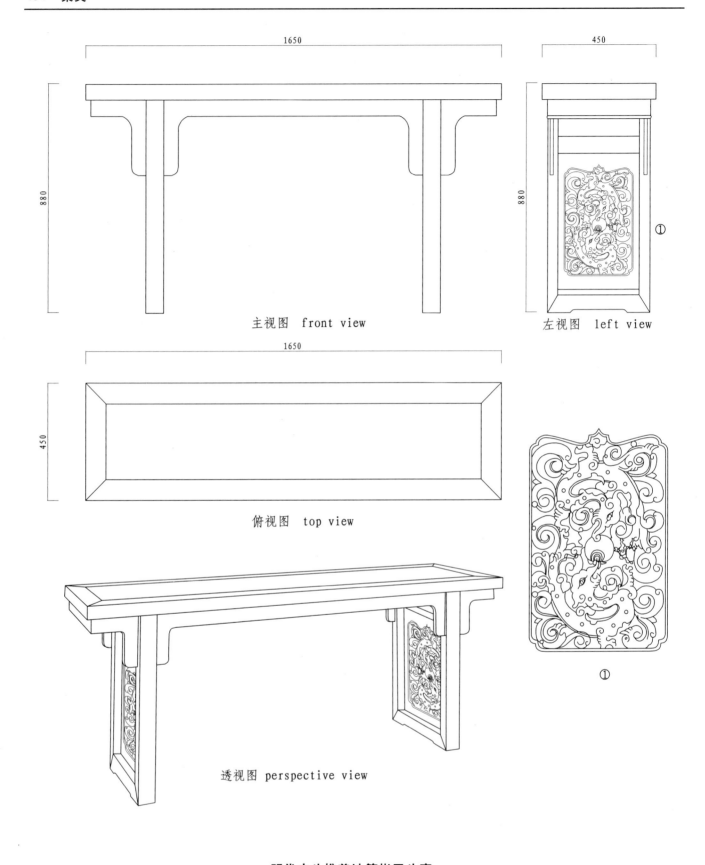

明代夹头榫着地管脚平头案
Ming dynasty flat-top narrow recessed-leg table with elongated bridle joints and base stretchers

明代黄花梨长方案
Ming dynasty *huanghuali* wood narrow rectangular recessed-leg table

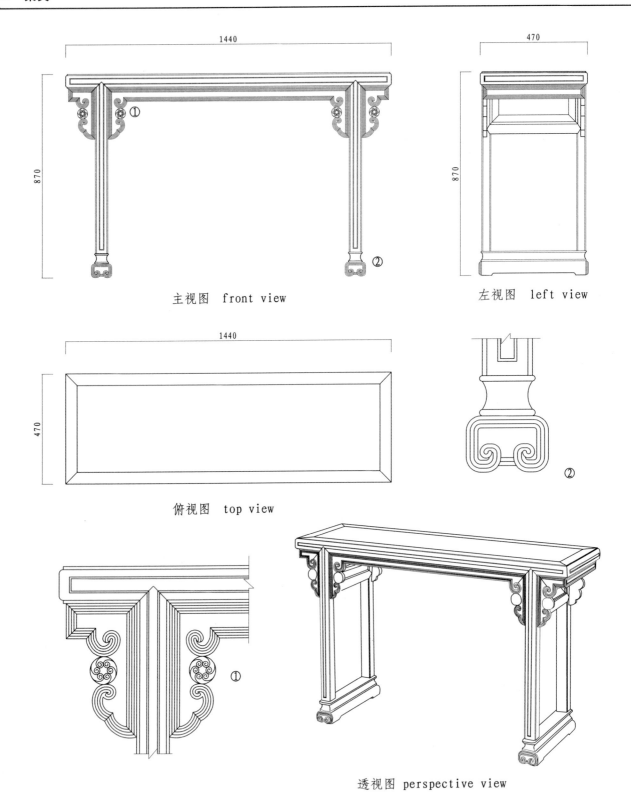

清代花梨卷云纹案
Qing dynasty *huali* wood narrow recessed-leg table with scroll-cloud pattern

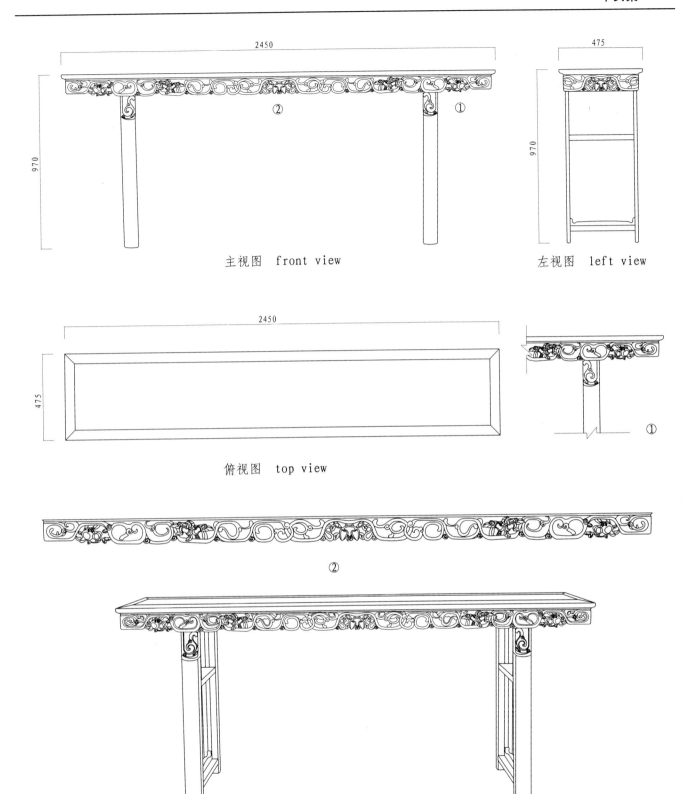

清代酸枝木卷草柿叶纹平头案
Qing dynasty *suanzhi* wood flat-top narrow recessed-leg table with curling tendril design and persimmon leaves pattern

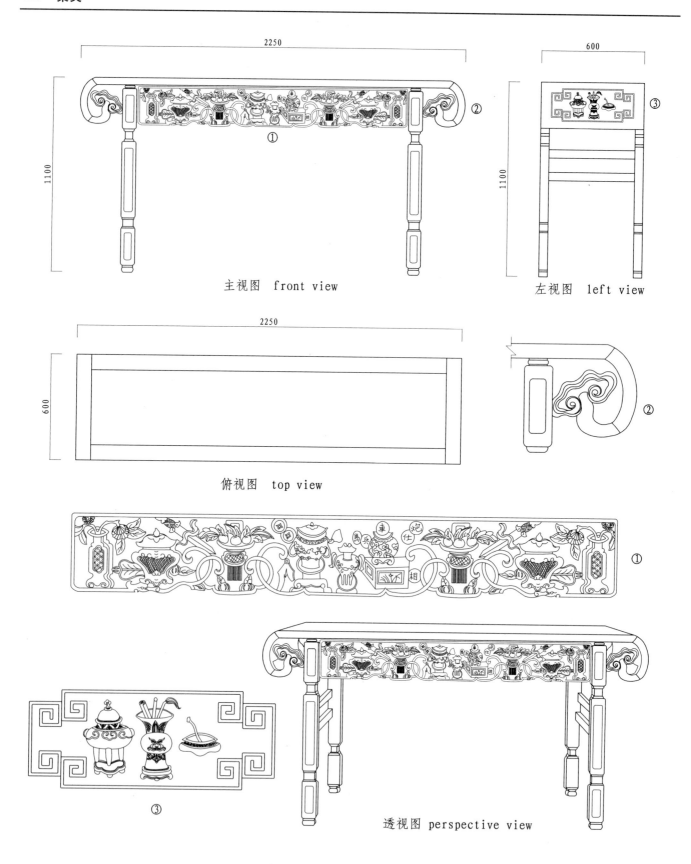

主视图 front view
左视图 left view
俯视图 top view
透视图 perspective view

清代花梨木博古纹卷头案
Qing dynasty *huali* wood narrow recessed-leg table with scroll termination and design of numerous antique and precious objects

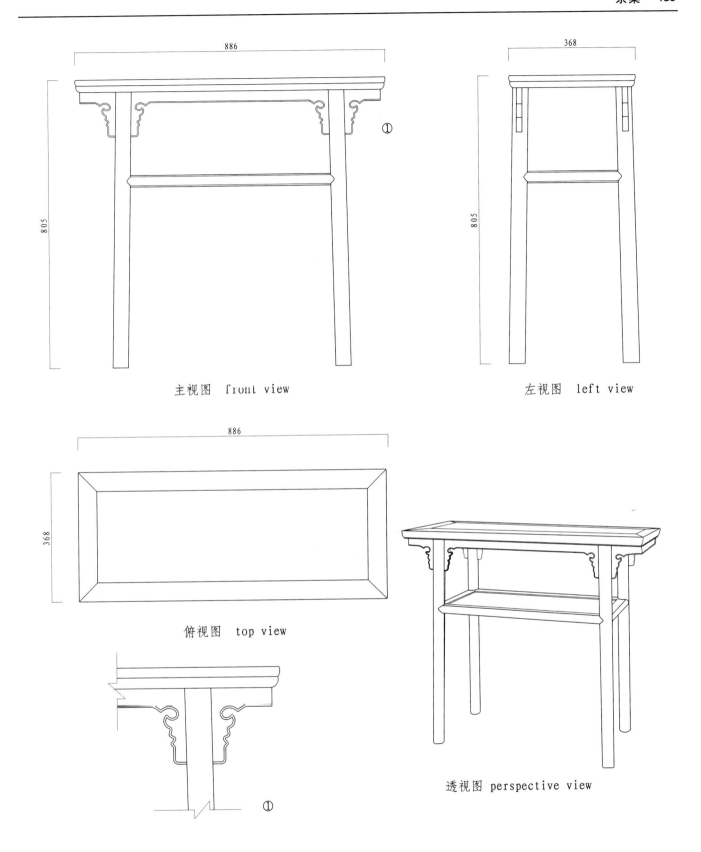

明代黄花梨夹头榫带屉板小条案
Ming dynasty *huanghuali* wood small narrow recessed-leg table with elongated bridle joints and a shelf panel

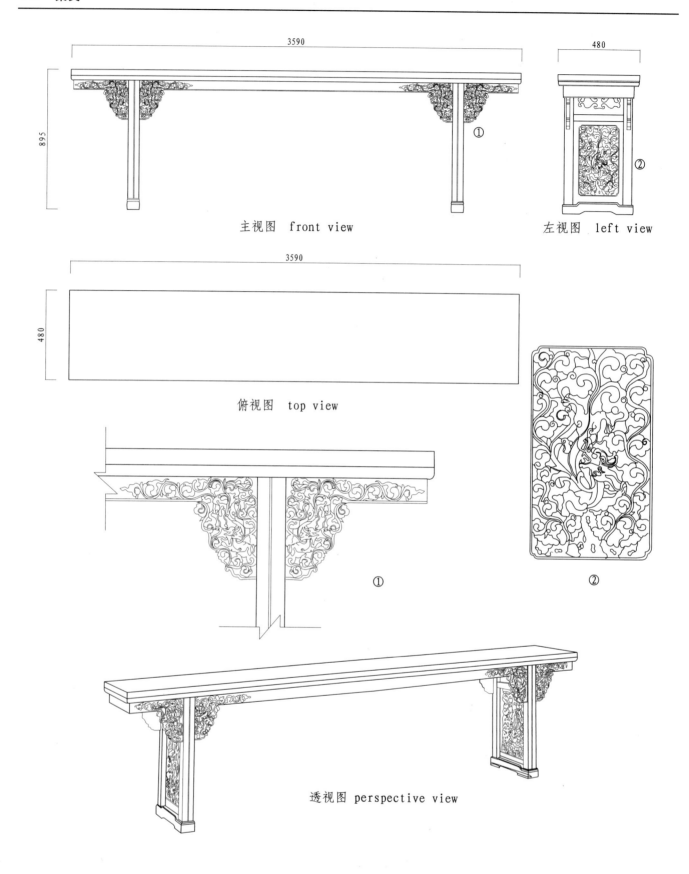

主视图 front view　　左视图 left view

俯视图 top view

① ②

透视图 perspective view

明代花梨云龙纹条案
Ming dynasty *huali* wood narrow recessed-leg table with cloud-and-dragon design

香几 141

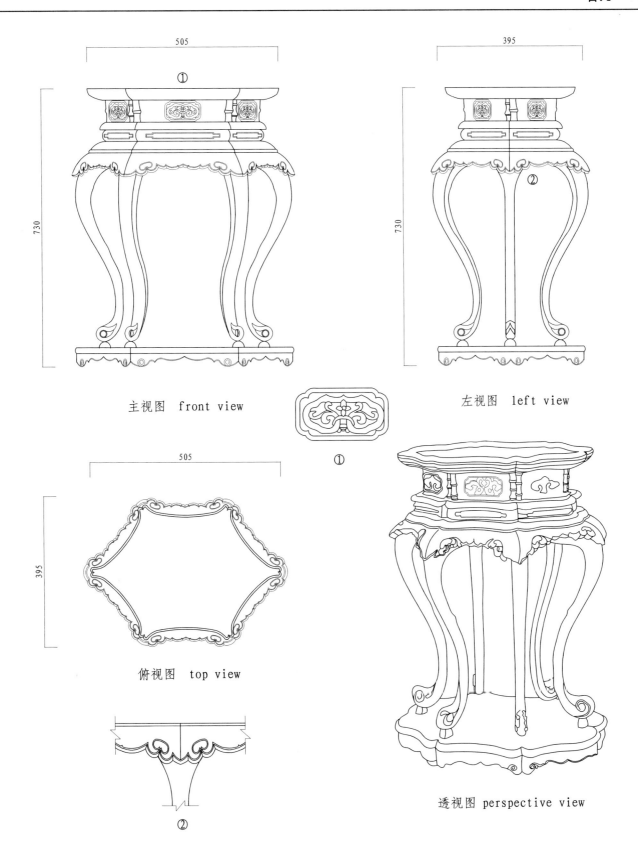

主视图 front view

左视图 left view

俯视图 top view

透视图 perspective view

明代黄花梨荷叶式六足香几
Ming dynasty *huanghuali* wood louts-leaf style incense stand with six legs

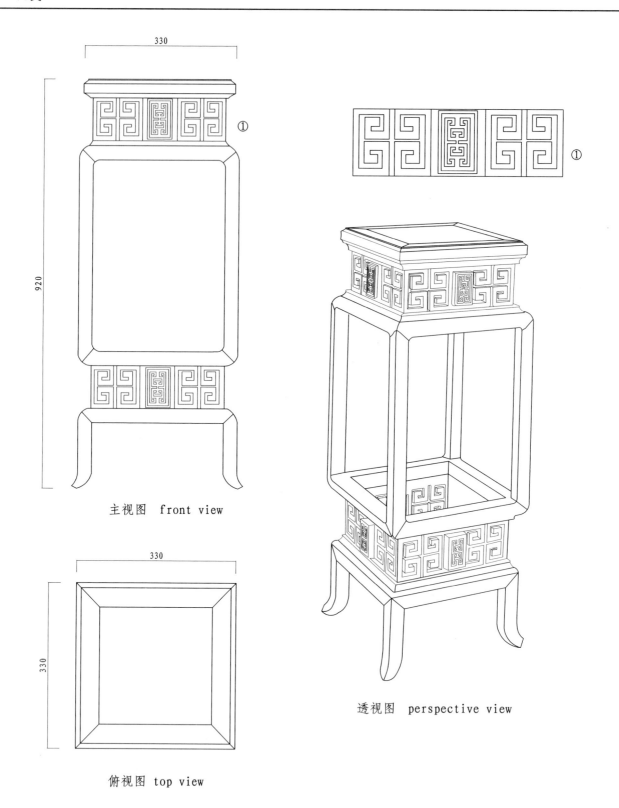

主视图 front view

俯视图 top view

透视图 perspective view

清代花梨回纹香几
Qing dynasty *huali* wood incense stand with rectangular spiral pattern

香几 143

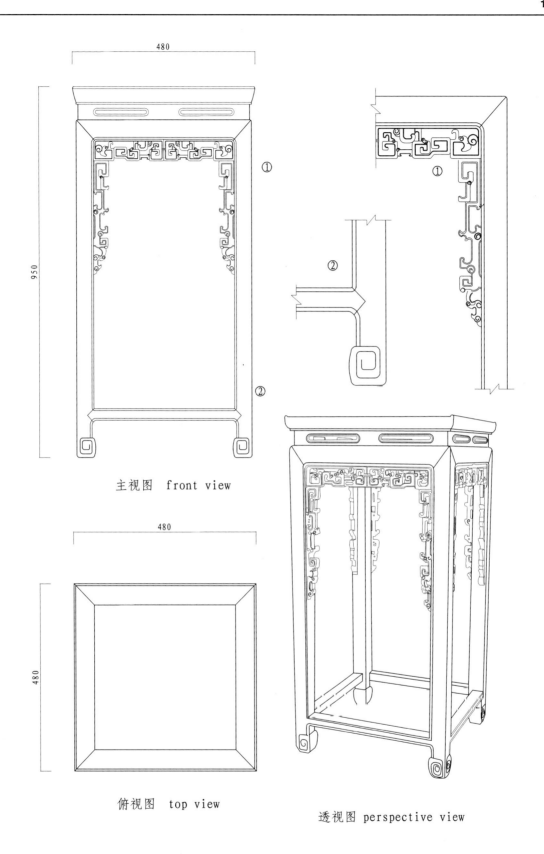

主视图 front view

俯视图 top view

透视图 perspective view

明代鸡翅木有束腰香几
Ming dynasty *jichi* wood waisted incense stand

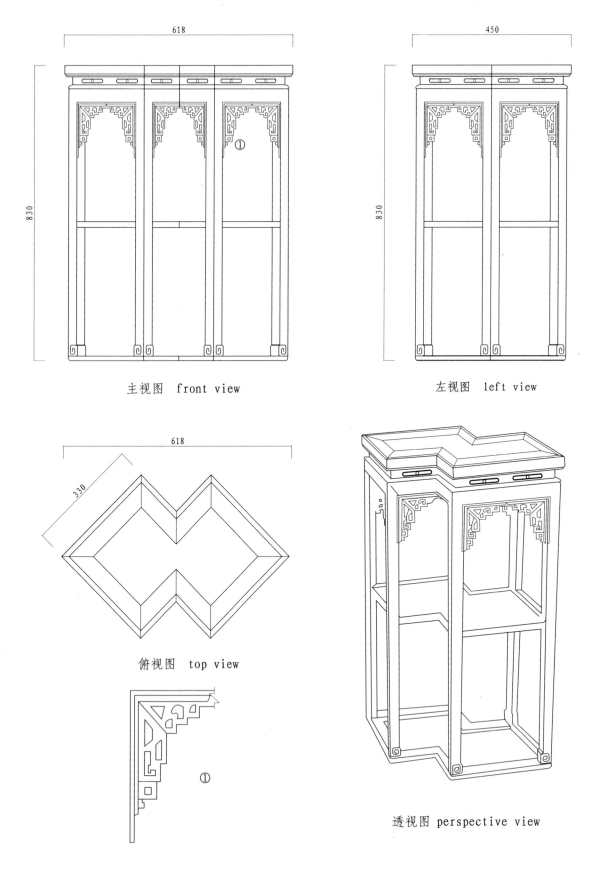

主视图 front view

左视图 left view

俯视图 top view

透视图 perspective view

清代红木方胜形香几
Qing dynasty *hong* wood incense stand with interlocked diamond design

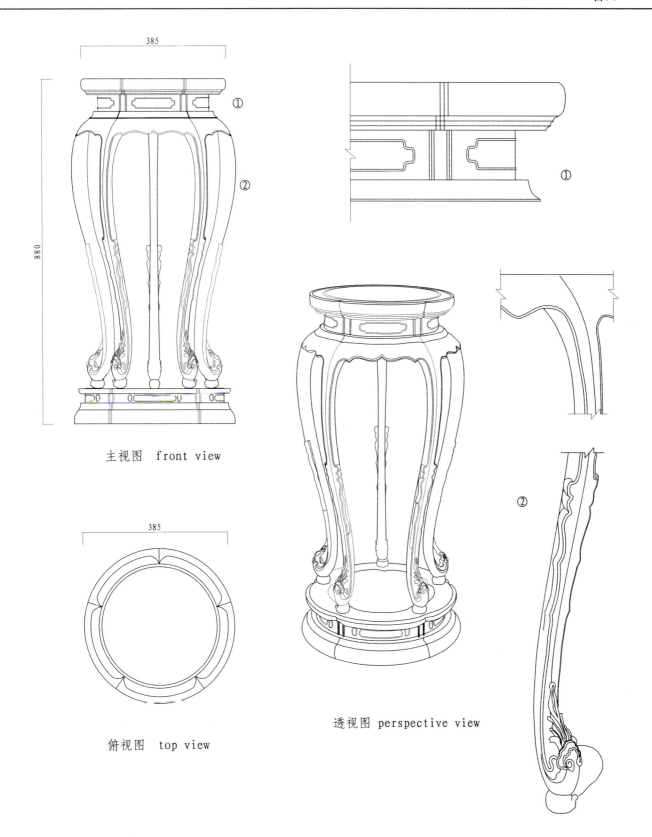

明代红漆嵌珐琅面梅花式香几
Ming dynasty plum-flower style incense stand with red lacquer and enamel inlay

146 几类

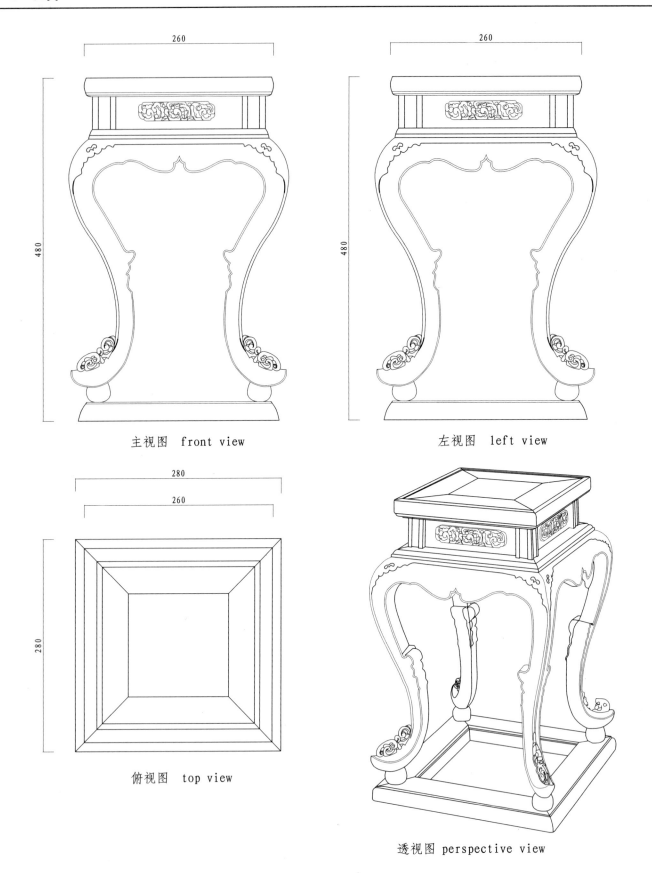

主视图 front view

左视图 left view

俯视图 top view

透视图 perspective view

明代黄花梨方香几
Ming dynasty *haunghuali* wood square incense stand

147　香几

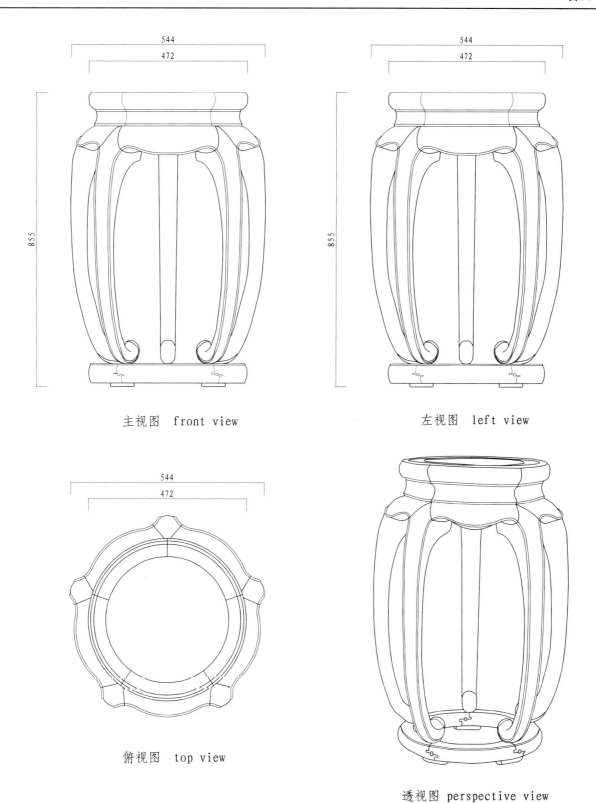

主视图 front view

左视图 left view

俯视图 top view

透视图 perspective view

明代红木五足内卷香几
Ming dynasty *hong* wood incense stand with five incurved legs

148　几类

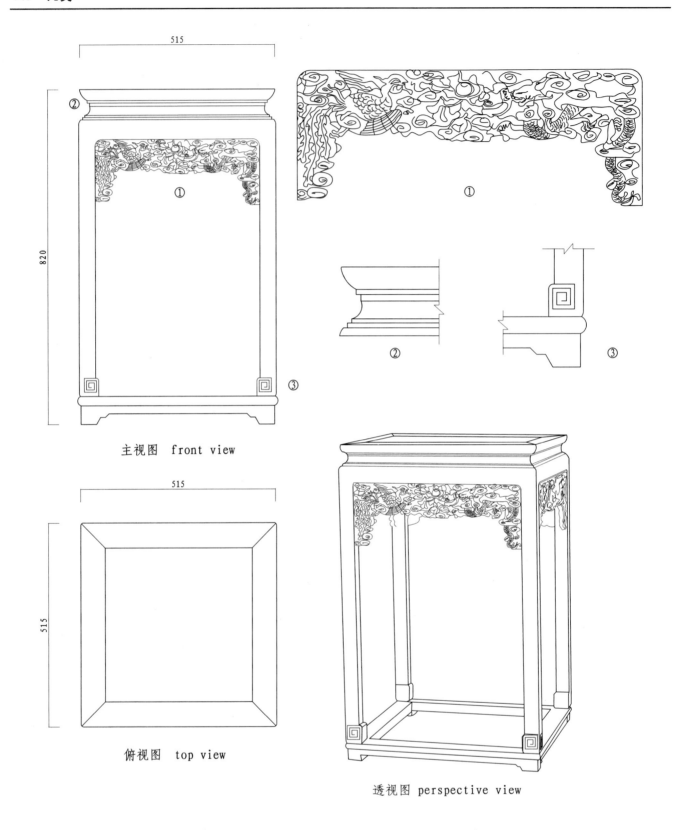

主视图　front view

俯视图　top view

透视图　perspective view

清乾隆紫檀木四方香几
Qing dynasty Qianlong period *zitan* wood square incense stand

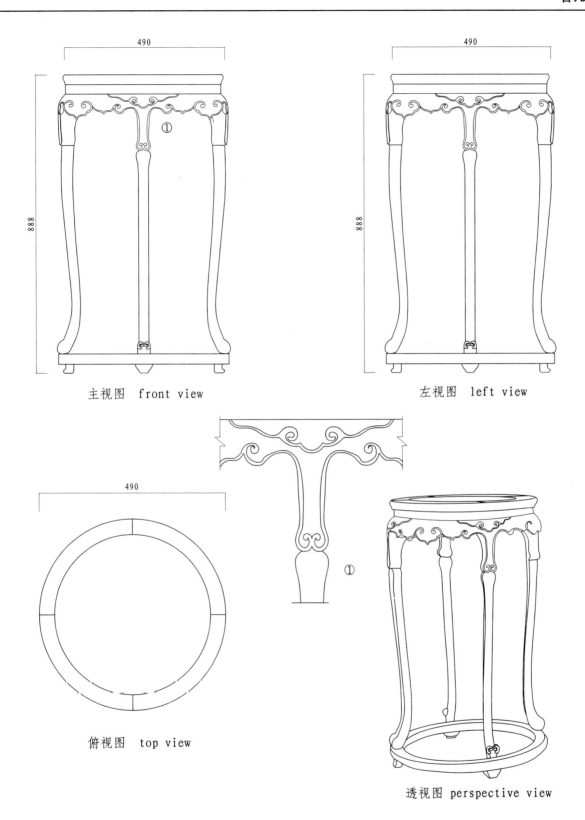

主视图 front view   左视图 left view

俯视图 top view   透视图 perspective view

明代老花梨木四足圆香几
Ming dynasty old *huali* wood round incense stand with four legs

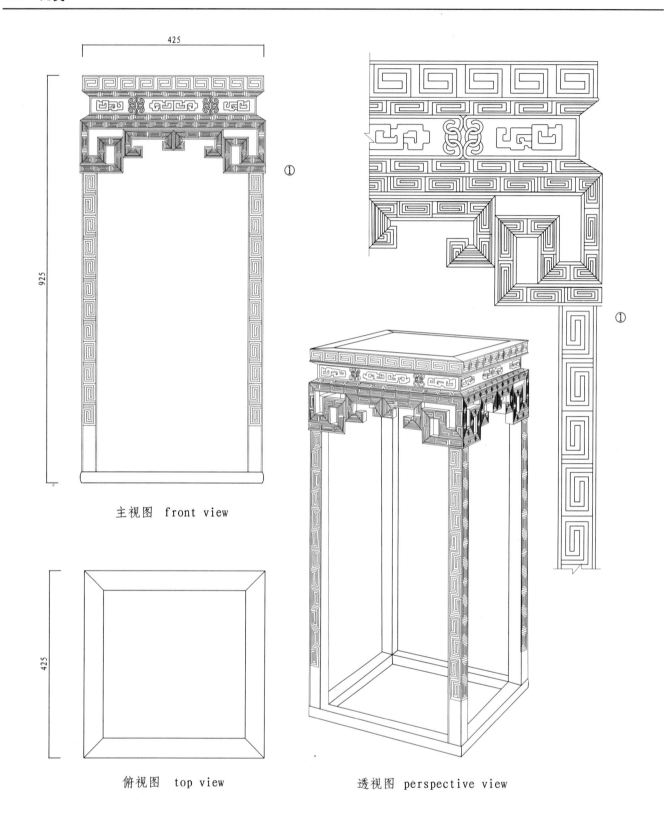

主视图 front view

俯视图 top view

透视图 perspective view

清乾隆楠木嵌竹丝回纹香几
Qing dynasty Qianlong period *nan* wood incense stand with bamboo filament inlay and rectangular spiral pattern

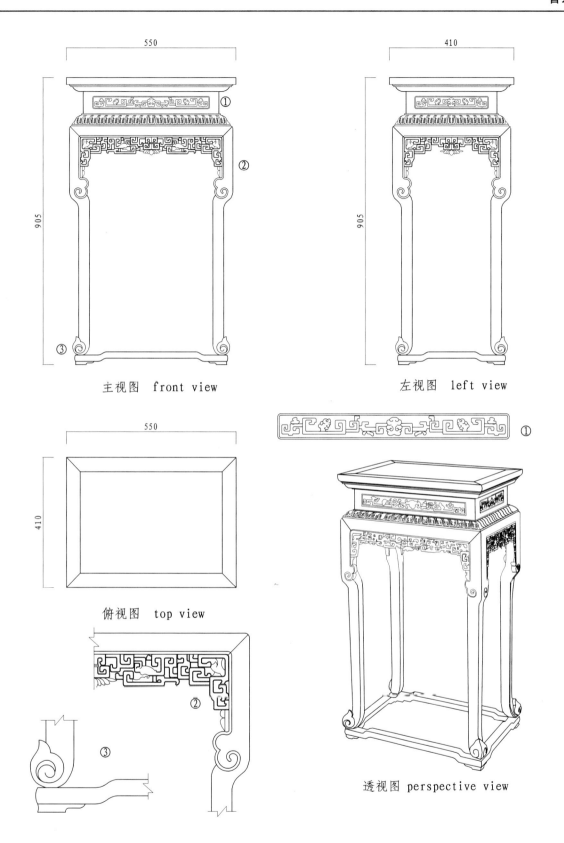

清乾隆紫檀夔龙纹香几
Qing dynasty Qianlong period *zitan* wood incense stand win *kui*-dragon design

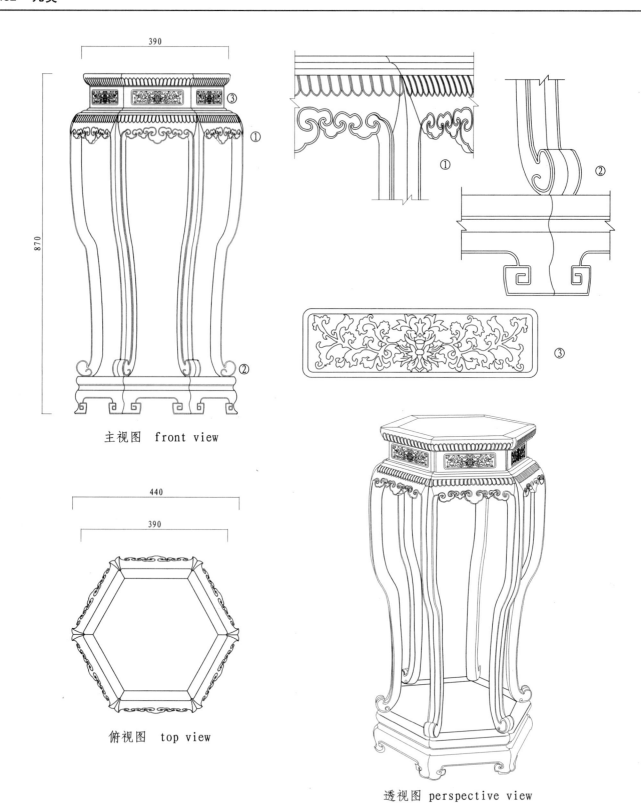

清乾隆紫檀西番莲纹六方香几
Qing dynasty Qianlong perion *zitan* wood hexagonal incense stand with passion flowers pattern

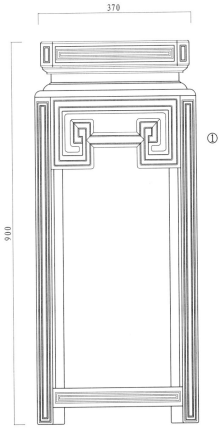

主视图 front view

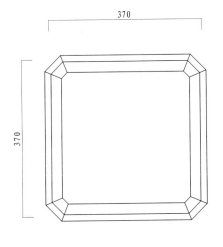

俯视图 top view

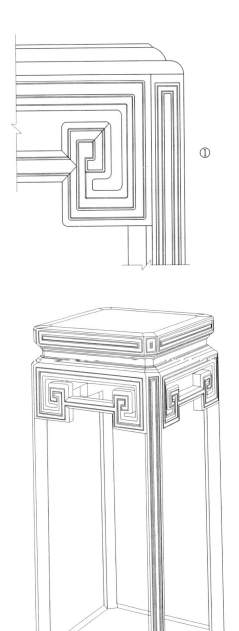

透视图 perspective view

清代鸡翅木镶紫檀回纹香几
Qing dynasty *jichi* wood incense stand with *zitan* wood inlay and rectangular spiral pattern

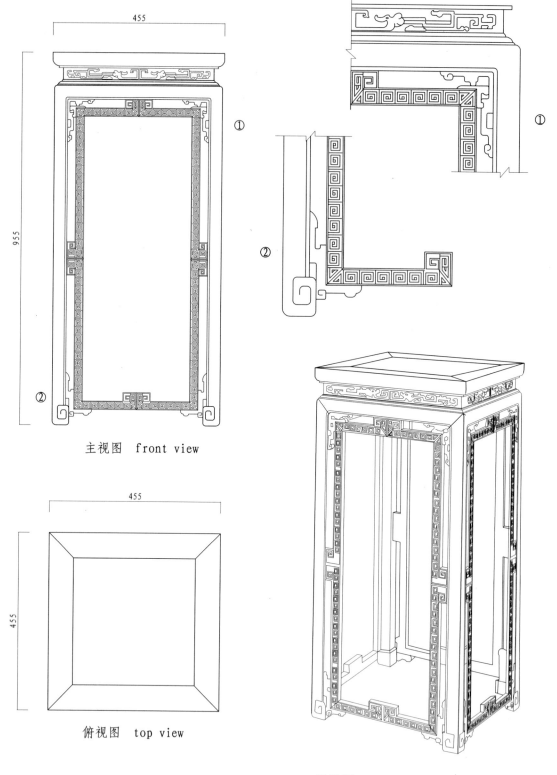

主视图 front view

俯视图 top view

透视图 perspective view

清代紫檀拐子纹香几
Qing dynasty *zitan* wood incense stand with rectangular spiral pattern

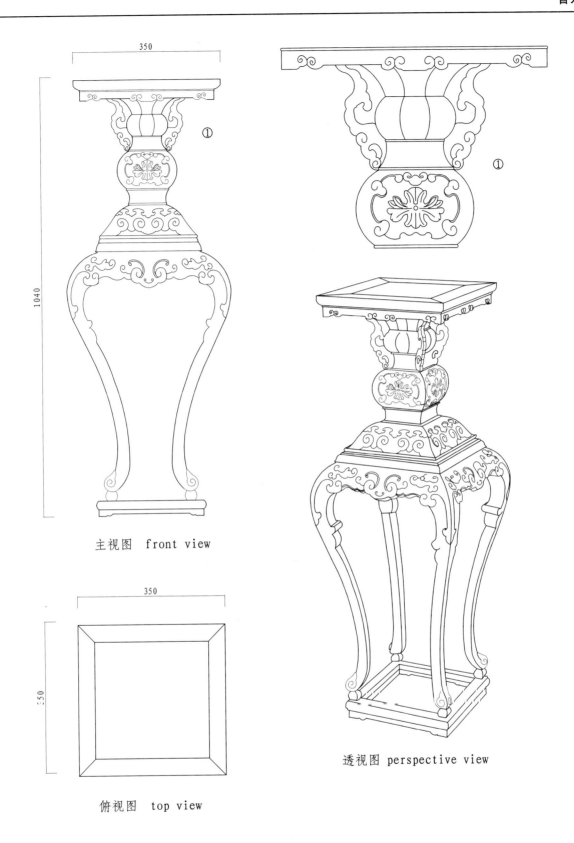

主视图 front view

俯视图 top view

透视图 perspective view

清代紫檀瓶式香几
Qing dynasty *zitan* wood vase-shaped incense stand

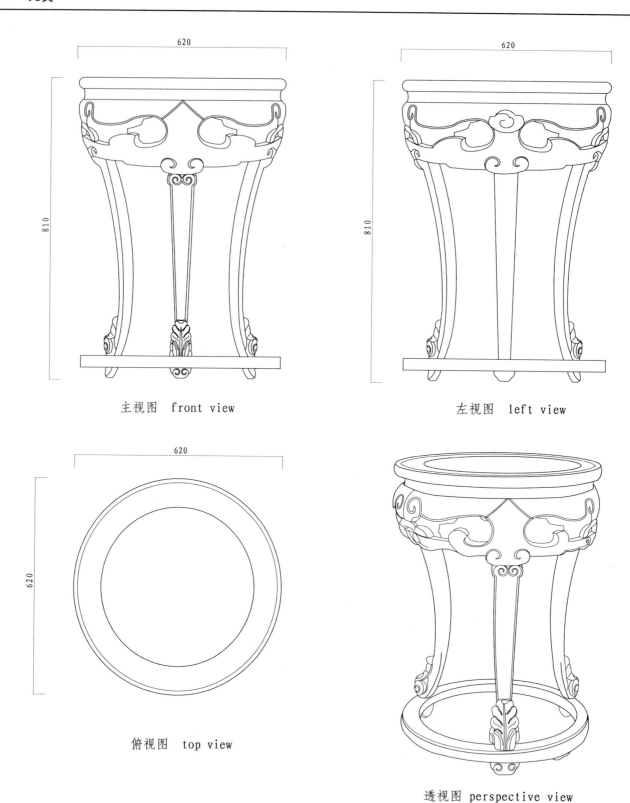

主视图 front view

左视图 left view

俯视图 top view

透视图 perspective view

清代紫檀三足圆香几
Qing dynasty *zitan* wood round incense stand with three legs

主视图 front view

俯视图 top view

透视图 perspective view

清代花梨木如意图花几
Qing dynasty *huali* wood flower stand with *ruyi* motif

主视图 front view

俯视图 top view

透视图 perspective view

清代束腰拐子龙花几
Qing dynasty waisted flower stand with rectangular spiral dragon design

主视图 front view

俯视图 top view

透视图 perspective view

清代拐子龙井口字花几
Qing dynasty flower stand with rectangular spiral dargon design and *jing* character pattern latticework

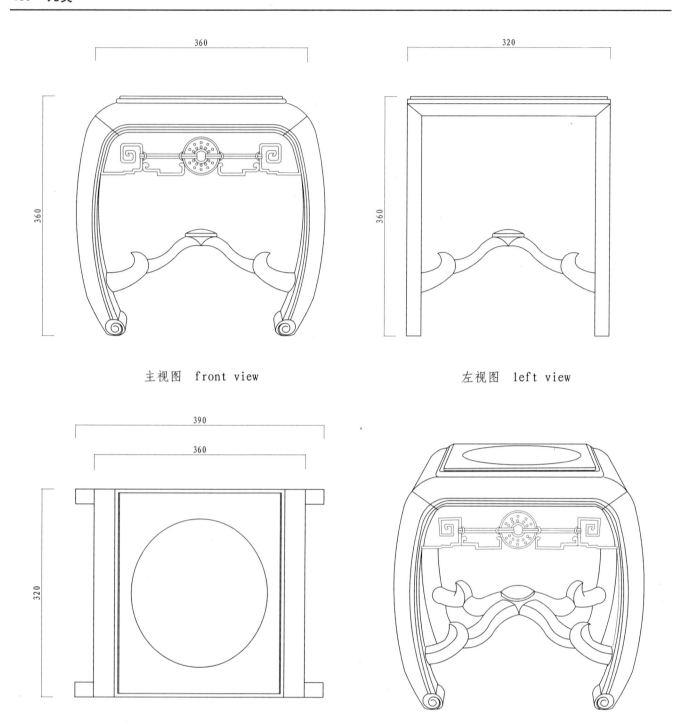

清代酸枝玉璧纹镶石小花几
Qing dynasty *suanzhi* wood small flower stand with marble inaly
and round flat pieces of jade design

茶几 161

主视图 front view

俯视图 top view

透视图 perspective view

清代紫檀木茶几
Qing dynasty *zitan* wood tea table

162　几类

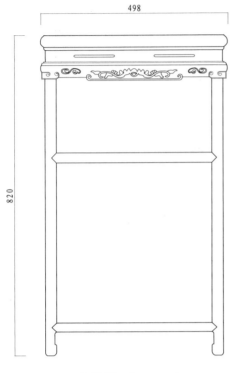

主视图　front view

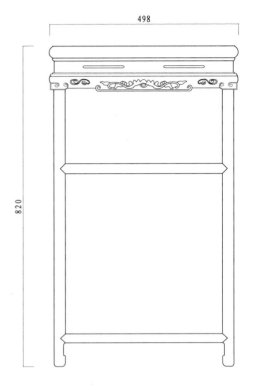

左视图　left view

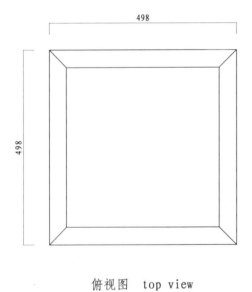

俯视图　top view

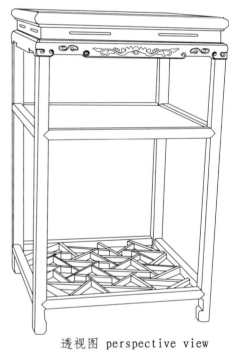

透视图　perspective view

清代红木方形茶几
Qing dynasty *hong* wood rectangular tea table

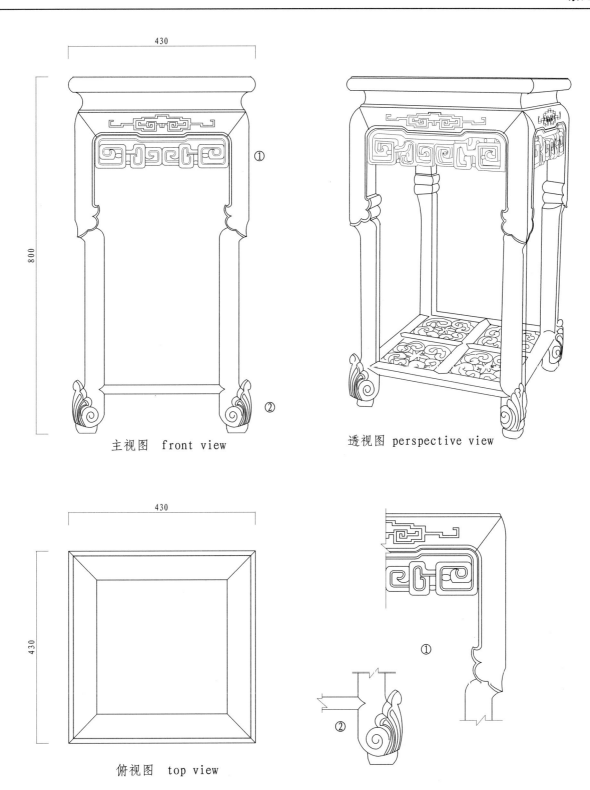

主视图 front view

透视图 perspective view

俯视图 top view

清代香红木茶几
Qing dynasty aromatic *hong* wood tea table

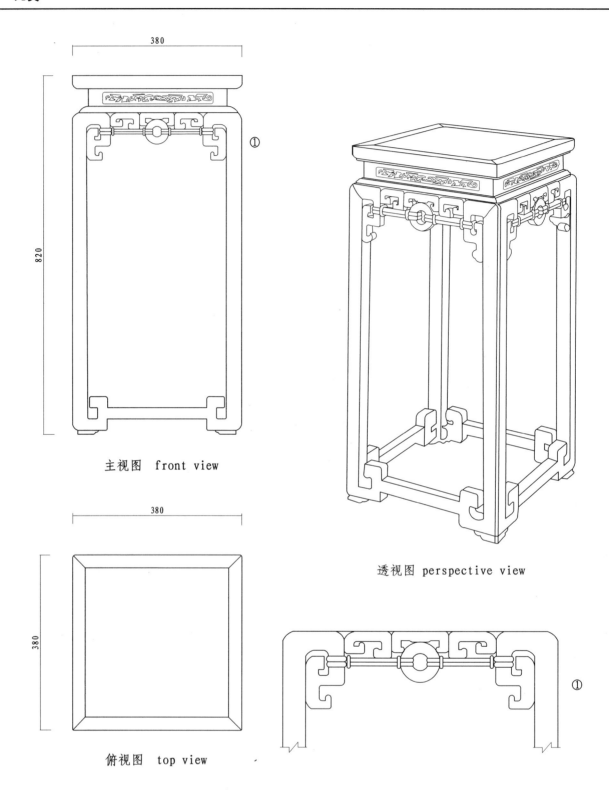

主视图 front view

俯视图 top view

透视图 perspective view

清代瘿木绳璧纹茶几
Qing dynasty burl wood tea table with rope-jade design

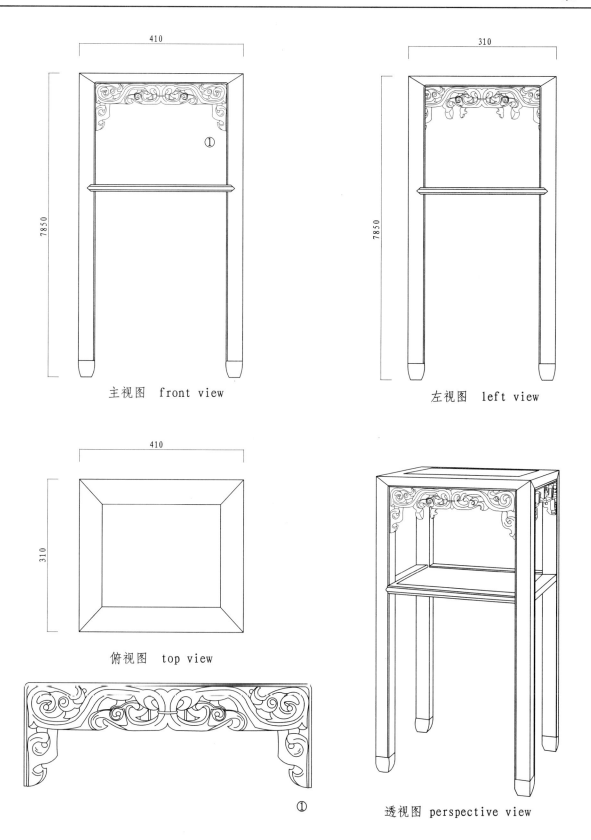

主视图 front view 左视图 left view 俯视图 top view 透视图 perspective view

清代酸枝木卷草纹茶几
Qing dynasty *suanzhi* wood tea table with curling tendril design

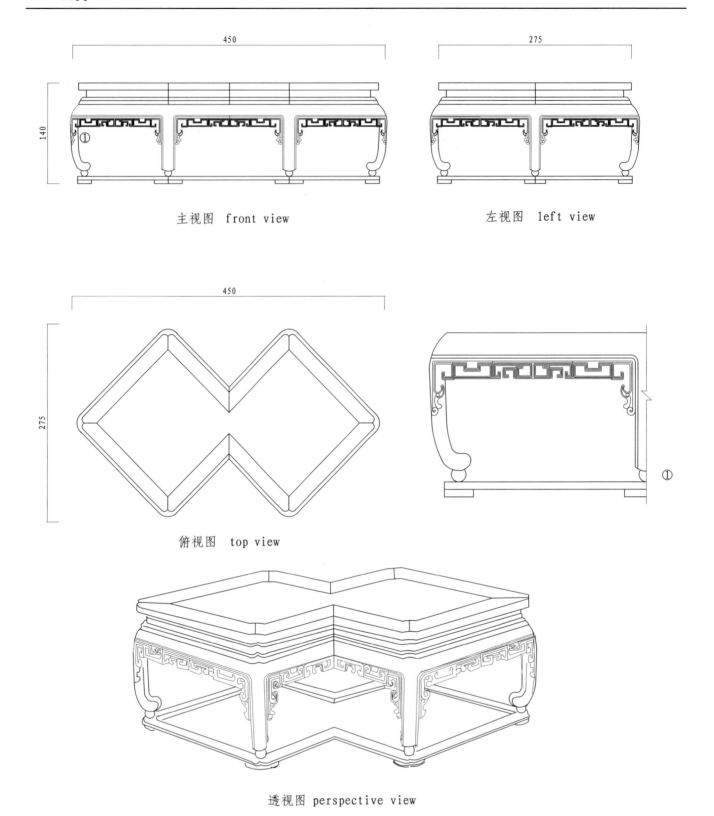

主视图 front view

左视图 left view

俯视图 top view

透视图 perspective view

清中期描金桃蝠纹方胜形几
Mid Qing dynasty interlocked-diamond-shaped table painting in gold with peach-and-bat pattern

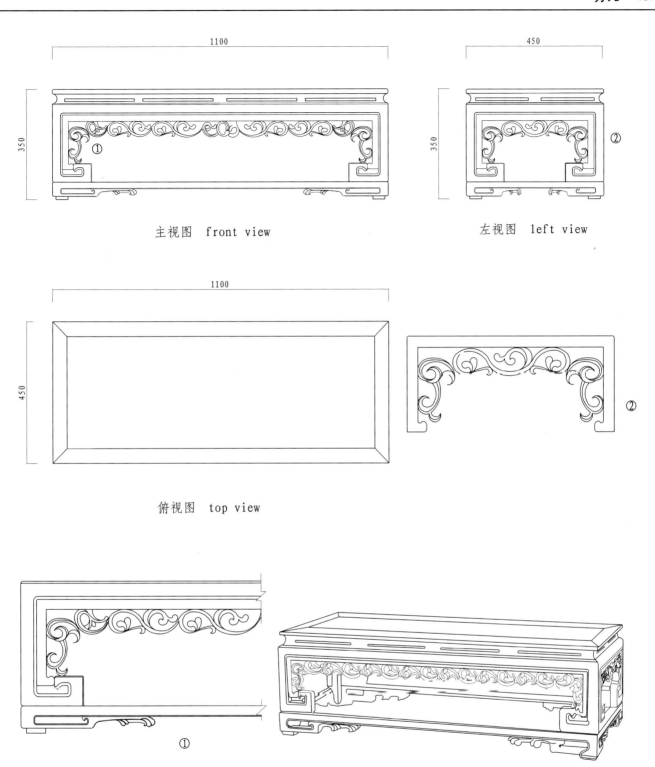

主视图 front view

左视图 left view

俯视图 top view

透视图 perspective view

清代紫檀长方小几
Qing dynasty *zitan* wood small rectangular table

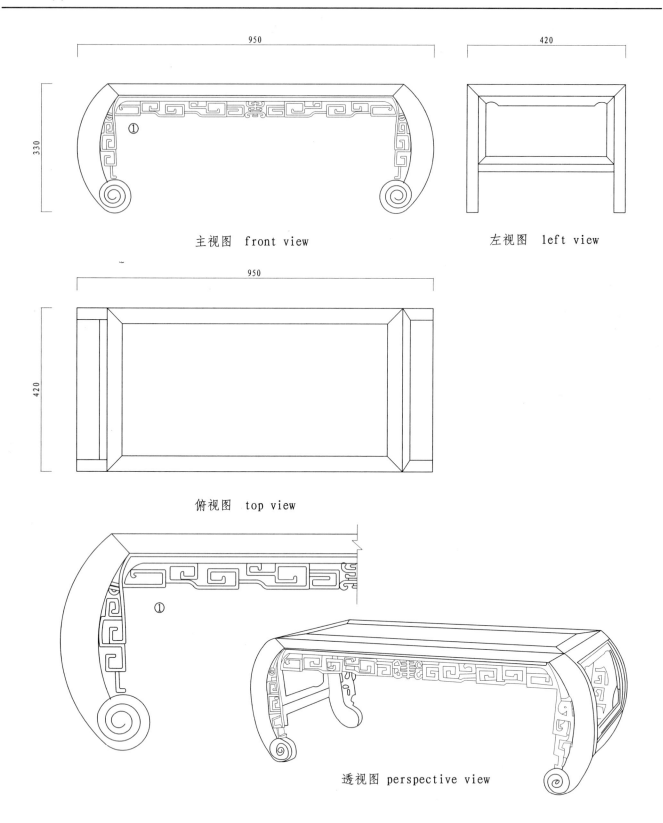

主视图 front view　　左视图 left view

俯视图 top view

透视图 perspective view

清代红木镶瘿木面炕几
Qing dynasty *hong* wood *kang* table with burl wood inlay

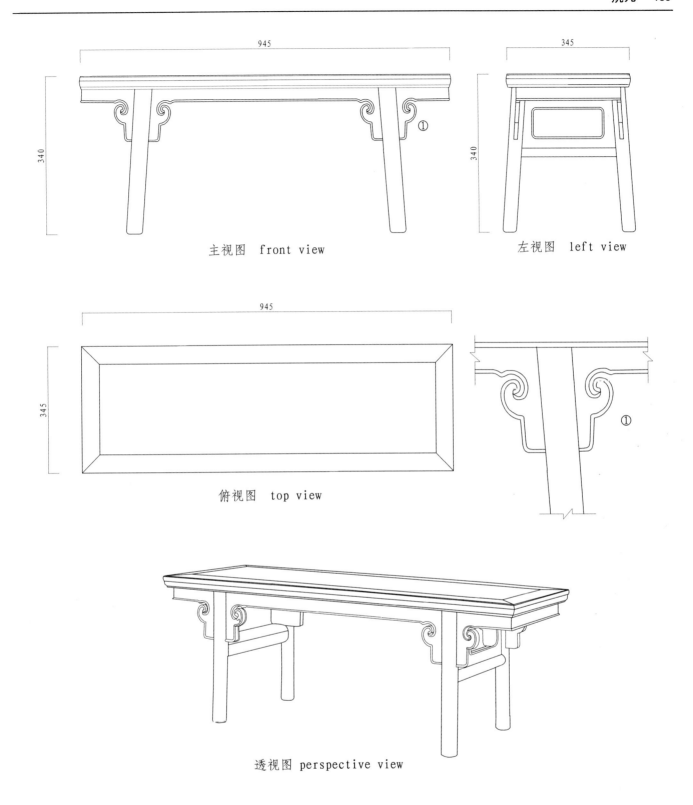

清早期紫檀炕几
Early Qing dynasty *zitan* wood *kang* table

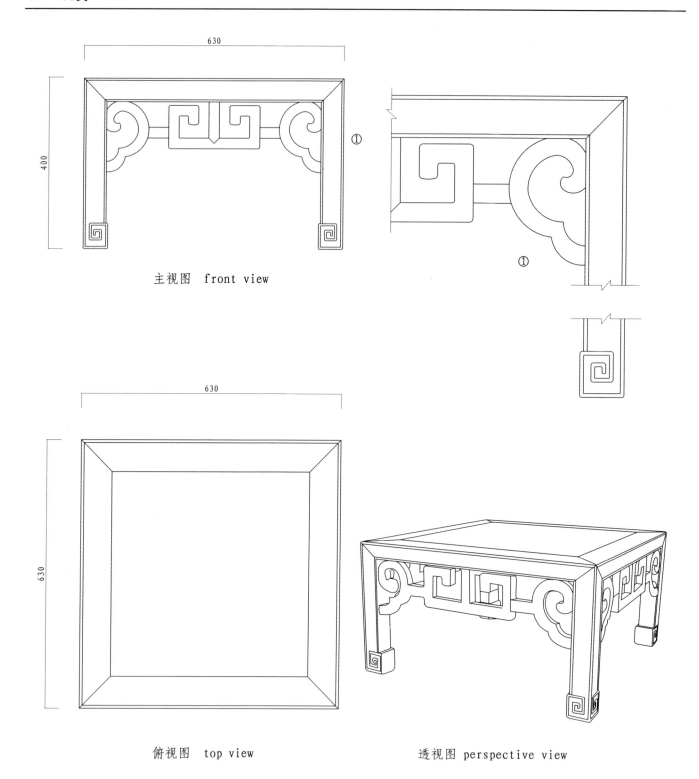

主视图 front view

俯视图 top view

透视图 perspective view

清代紫檀方炕几
Qing dynasty *zitan* wood square *kang* table

炕几 171

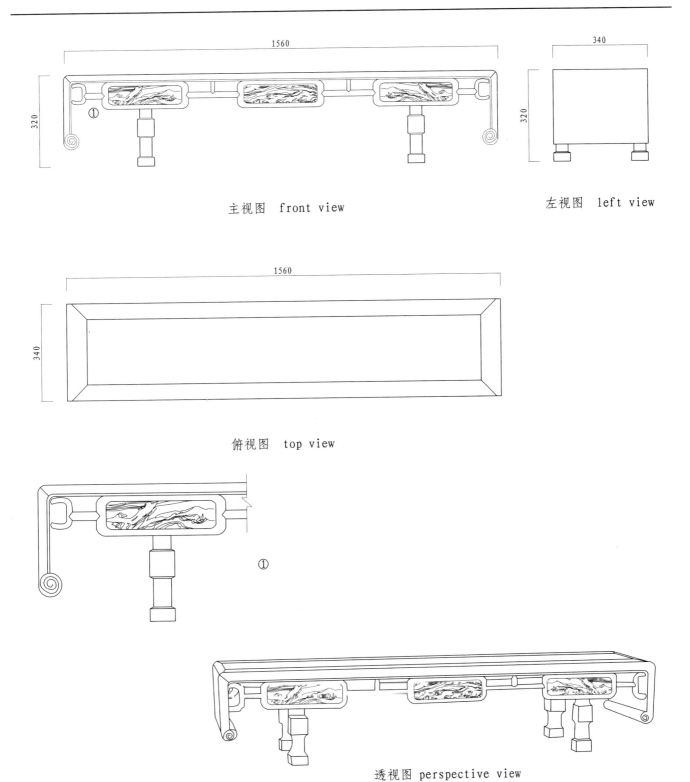

主视图 front view　　左视图 left view

俯视图 top view

透视图 perspective view

清代红木镶大理石炕几
Qing dynasty *hong* wood *kang* table with marble inlay

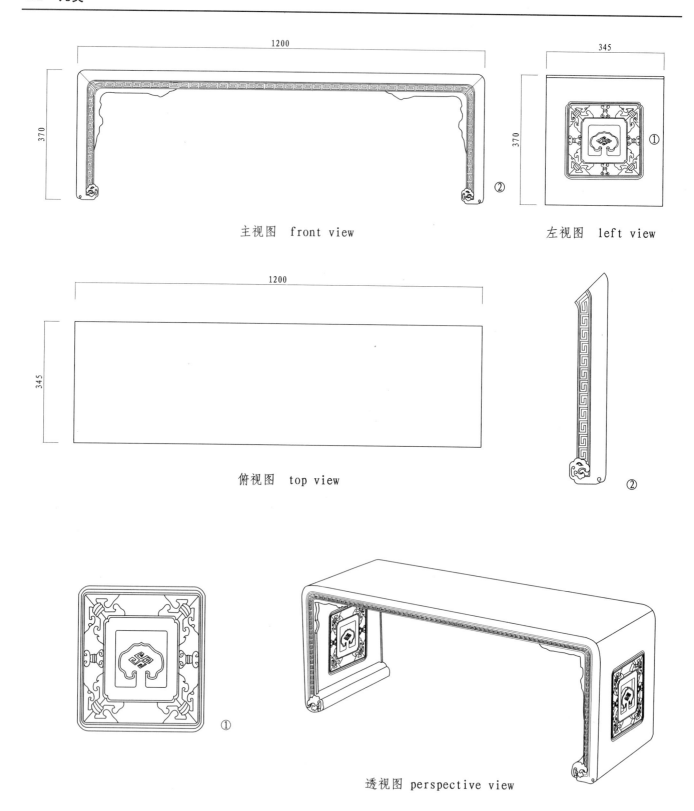

明代黄花梨透雕云纹炕几
Ming dynasty *huanghuali* wood narrow *kang* table with openwork carving of cloud design

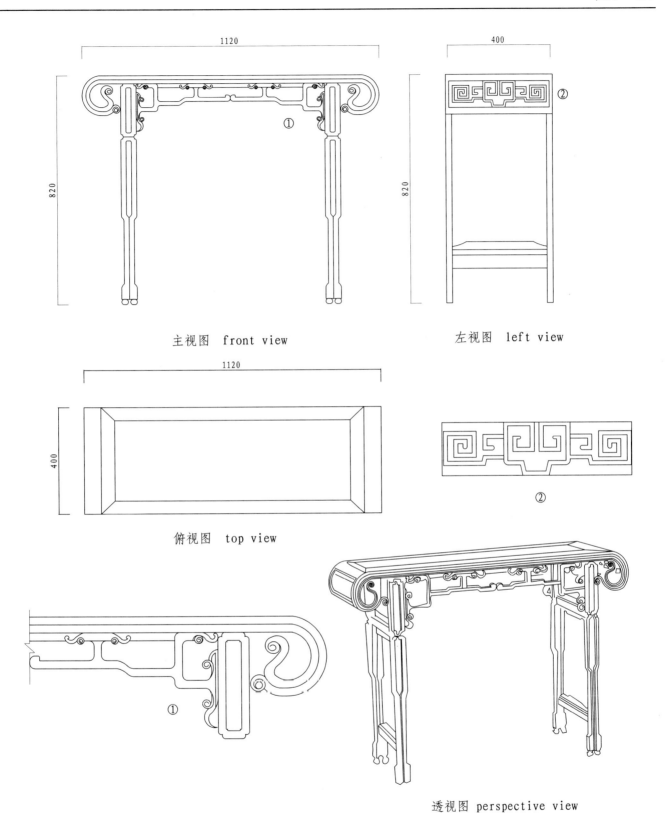

主视图 front view

左视图 left view

俯视图 top view

透视图 perspective view

清代酸枝书卷式花樽脚长几
Qing dynasty *suanzhi* wood narrow rectangular table with cylindrical scroll design and flower-goblet-shaped feet

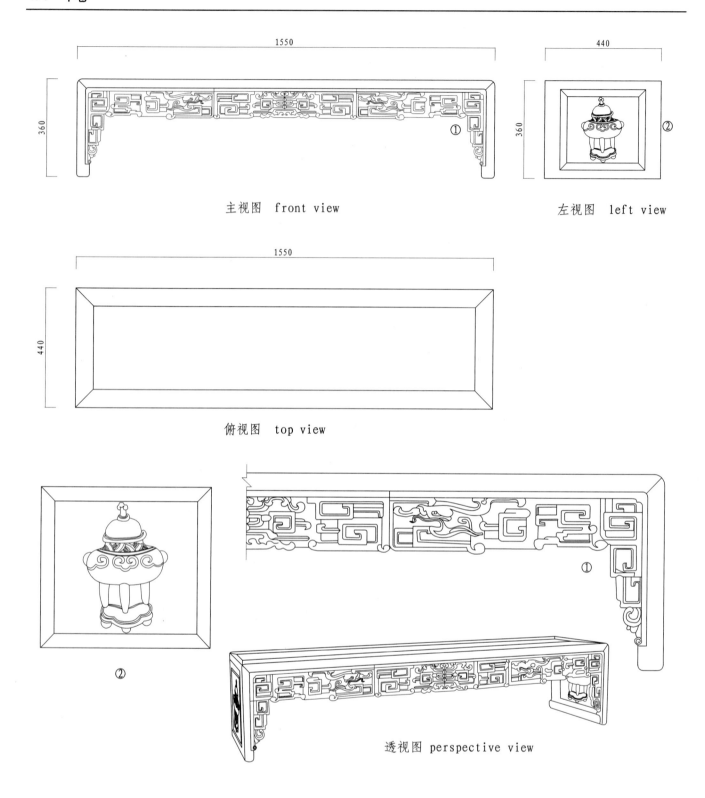

主视图 front view

左视图 left view

俯视图 top view

透视图 perspective view

清代核桃木拐子龙下卷
Qing dynasty walnut wood narrow rectangular table with open carving on the solid board legs and rectangular spiral dragon design

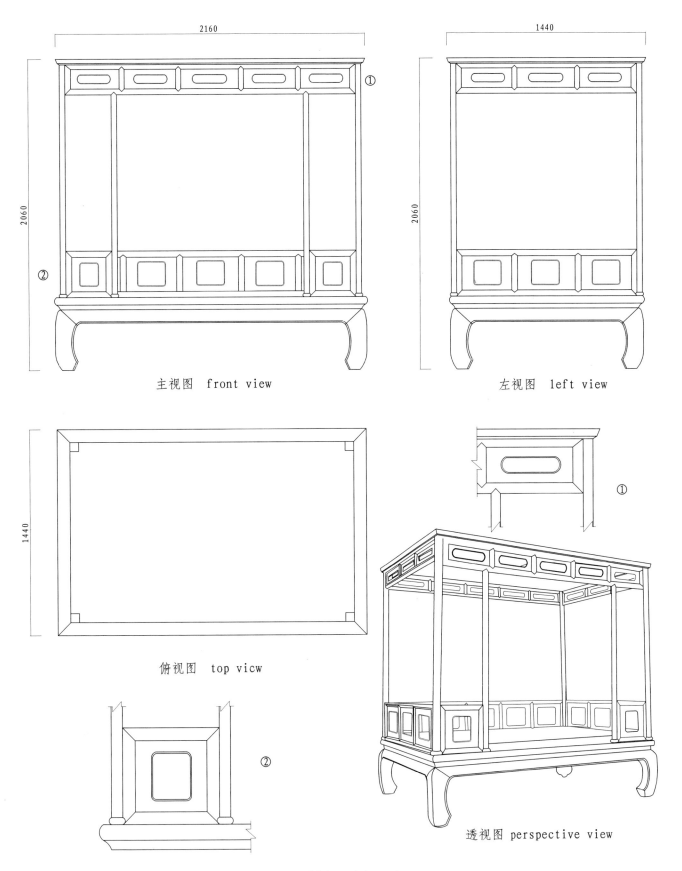

主视图 front view

左视图 left view

俯视图 top view

透视图 perspective view

明代榉木开光架子床
Ming dynasty *ju* wood canopy bed with medallion railings

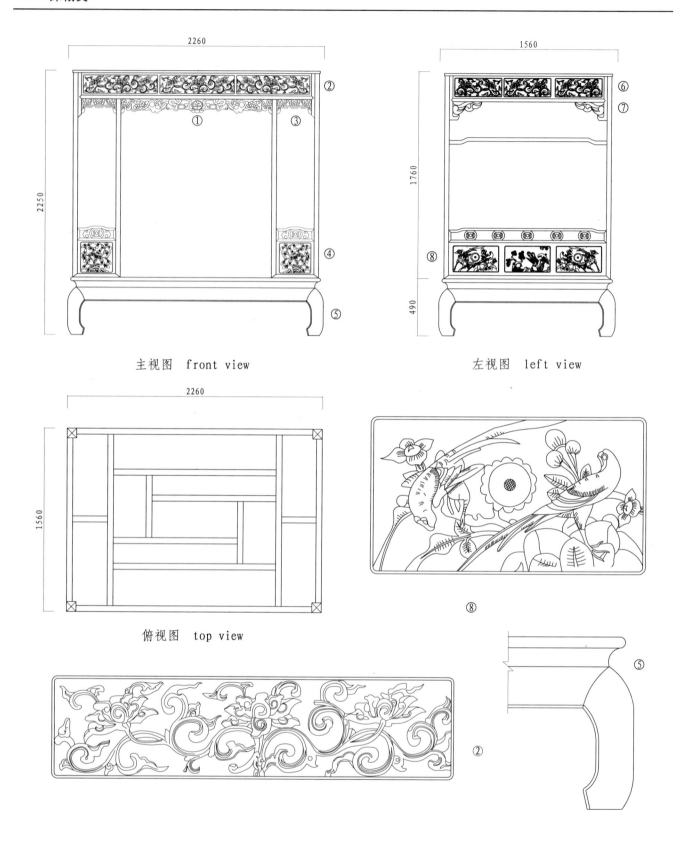

主视图 front view　　左视图 left view

俯视图 top view

清代紫檀红木合料八柱架子床
Qing dynasty canopy bed with eight posts made of *zitan* wood and *hong* wood

清代紫檀红木合料八柱架子床
Qing dynasty canopy bed with eight posts made of *zitan* wood and *hong* wood

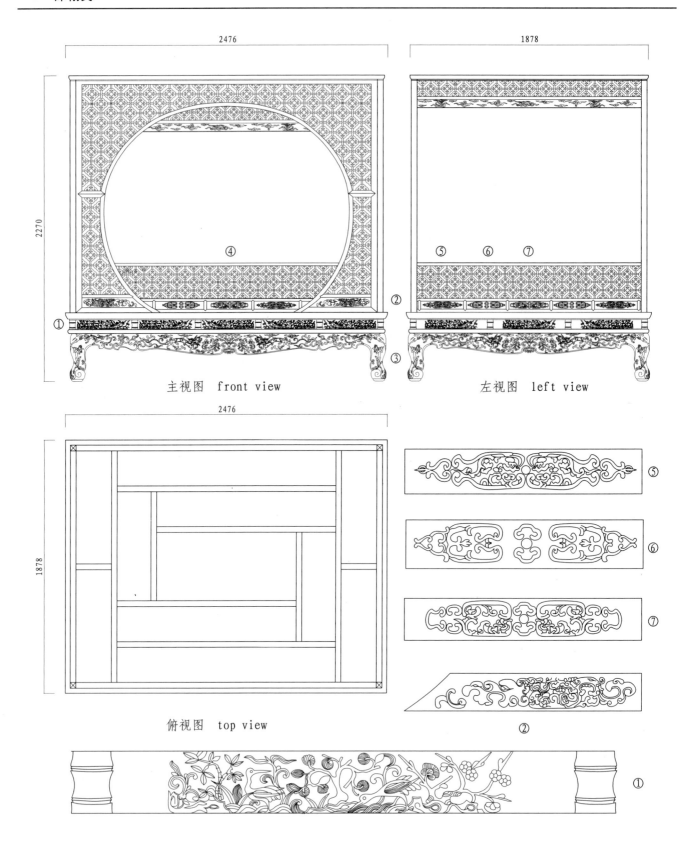

主视图 front view 左视图 left view 俯视图 top view

明代黄花梨月洞式门架子床
Ming dynasty *huanghuali* wood canopy bed with full-moon opening

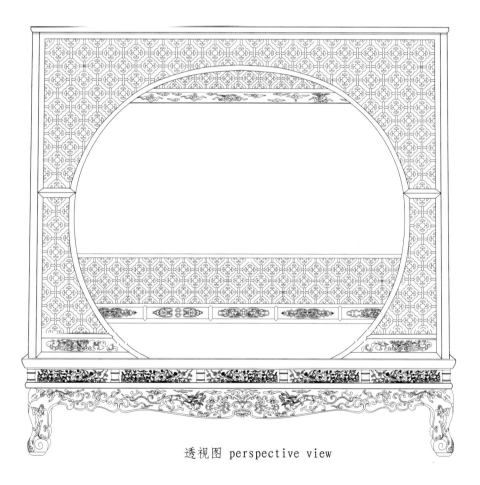

透视图 perspective view

明代黄花梨月洞式门架子床
Ming dynasty *huanghuali* wood canopy bed with full-moon opening

180 床榻类

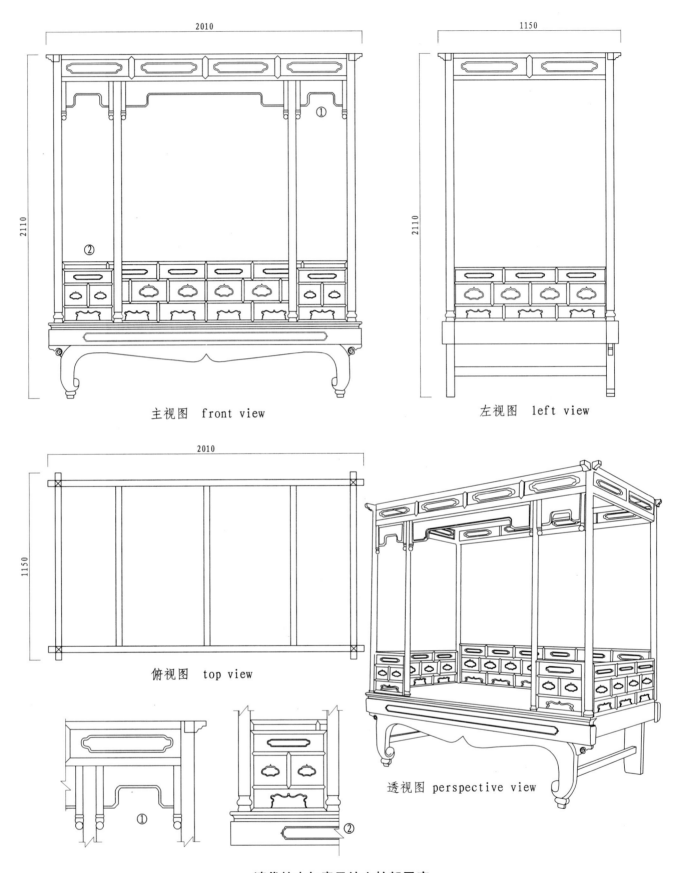

主视图 front view

左视图 left view

俯视图 top view

透视图 perspective view

清代柏木如意云纹六柱架子床
Qing dynasty cypress wood canopy bed with six posts and *ruyi*-cloud motif

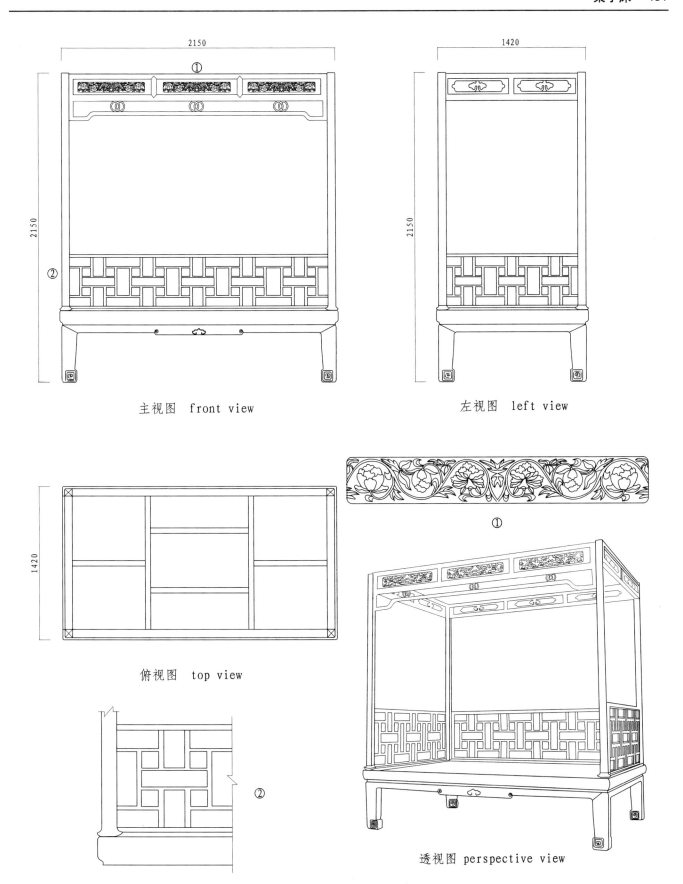

明代榉木直棂四柱架子床
Ming dynasty *ju* wood canopy bed with four posts and vertical rods

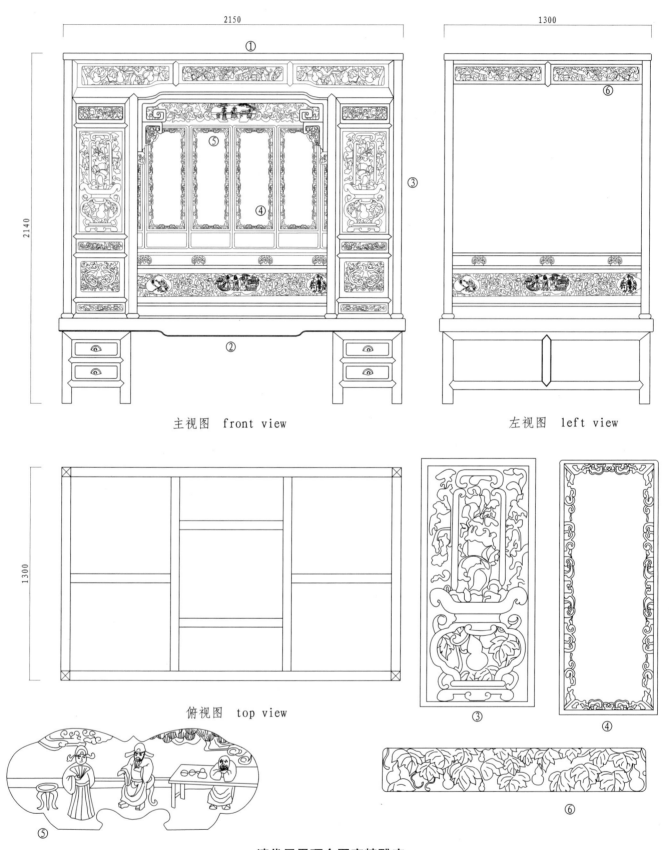

清代凤凰顶全围窗精雕床
Qing dynasty exquisite carving canopy bed with phoenix design top, largely enclosed railings and windows

①

②

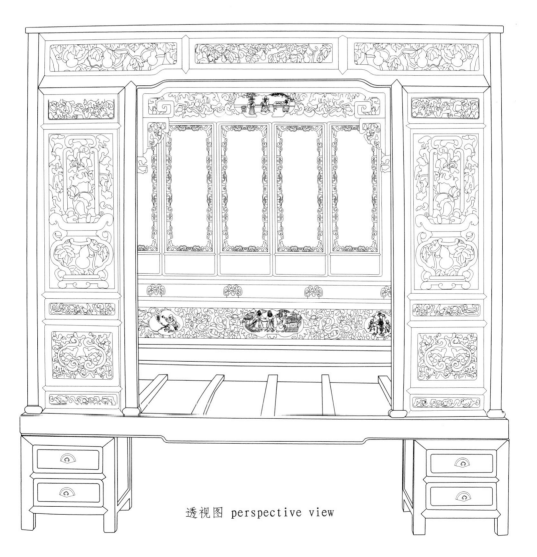

透视图 perspective view

清代凤凰顶全围窗精雕床
Qing dynasty exquisite carving canopy bed with phoenix design top, largely enclosed railings and windows

184 床榻类

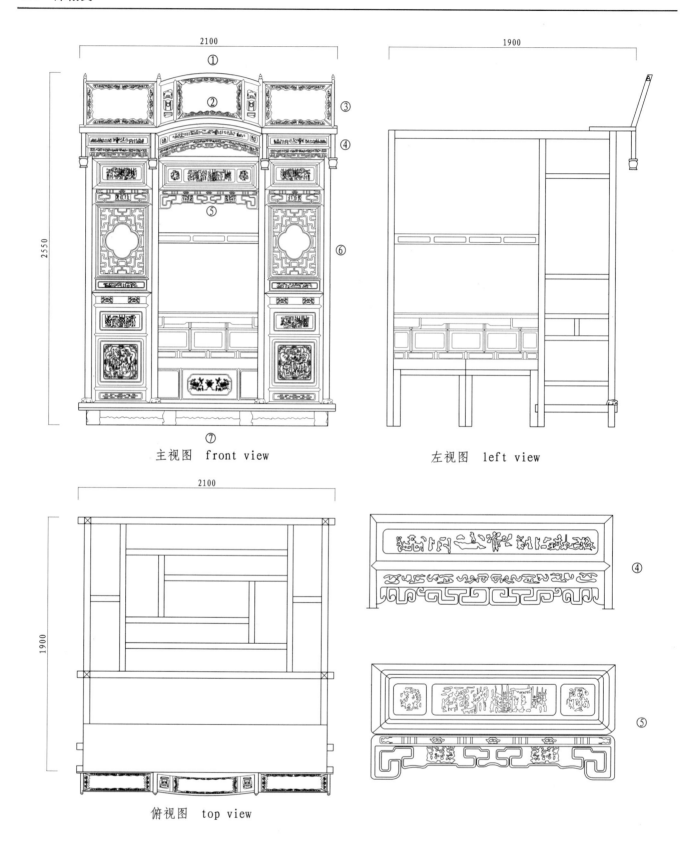

主视图 front view

左视图 left view

俯视图 top view

清代屋檐式嵌骨拔步床
Qing dynasty eave-shaped alcove bedsteat with bone inlay

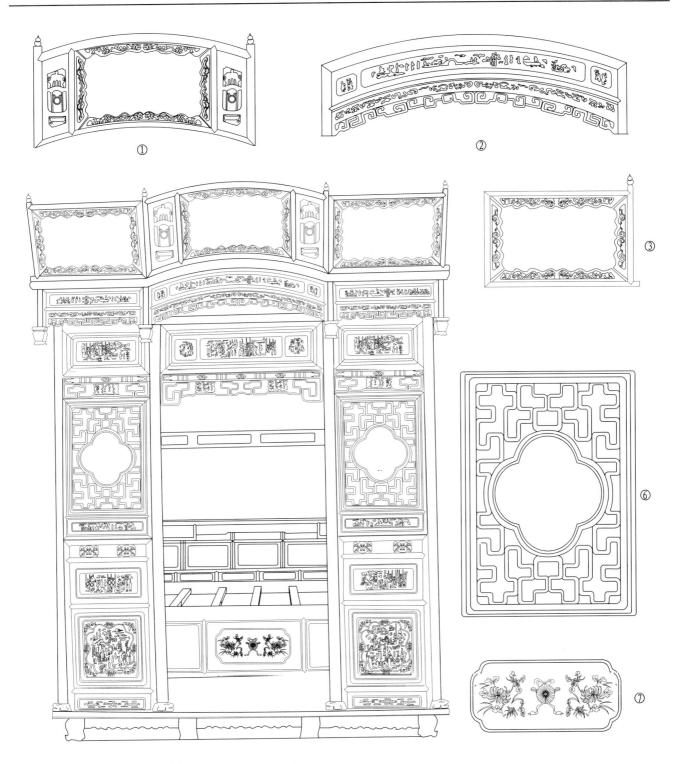

透视图 perspective view

清代屋檐式嵌骨拔步床
Qing dynasty eave-shaped alcove bedsteat with bone inlay

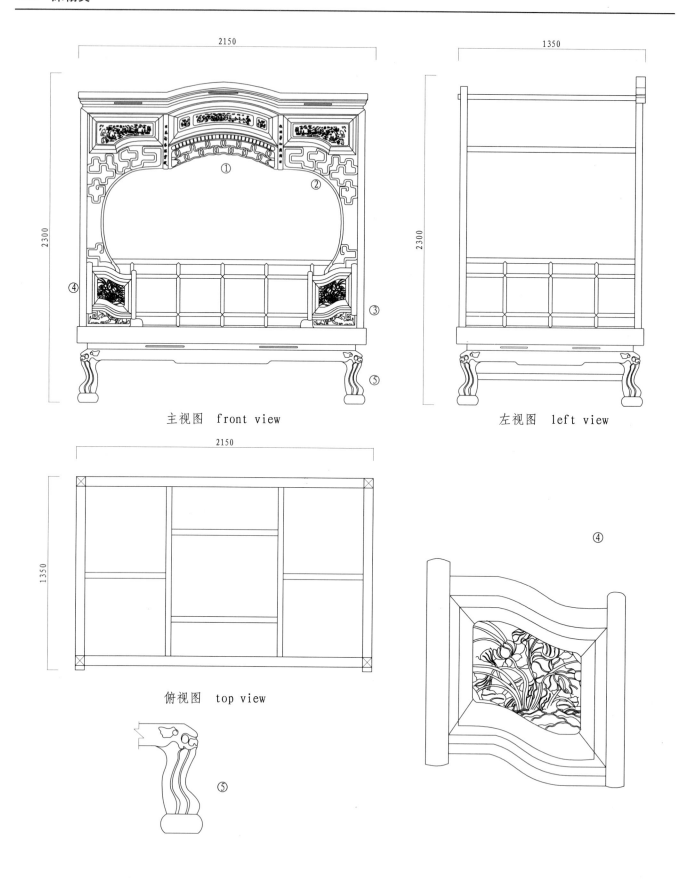

清代宁式柏木楹联床
Qing dynasty Ningbo style cypress wood bed with couplet design

透视图 perspective view

清代宁式柏木楹联床
Qing dynasty Ningbo style cypress wood bed with couplet design

主视图 front view

左视图 left view

俯视图 top view

清代一品爵位大开门红木床
Qing dynasty *hong* wood bed with wide opening for first rank of nobilities and officials

架子床 189

透视图 perspective view

清代一品爵位大开门红木床
Qing dynasty *hong* wood bed with wide opening for first rank of nobilities and officials

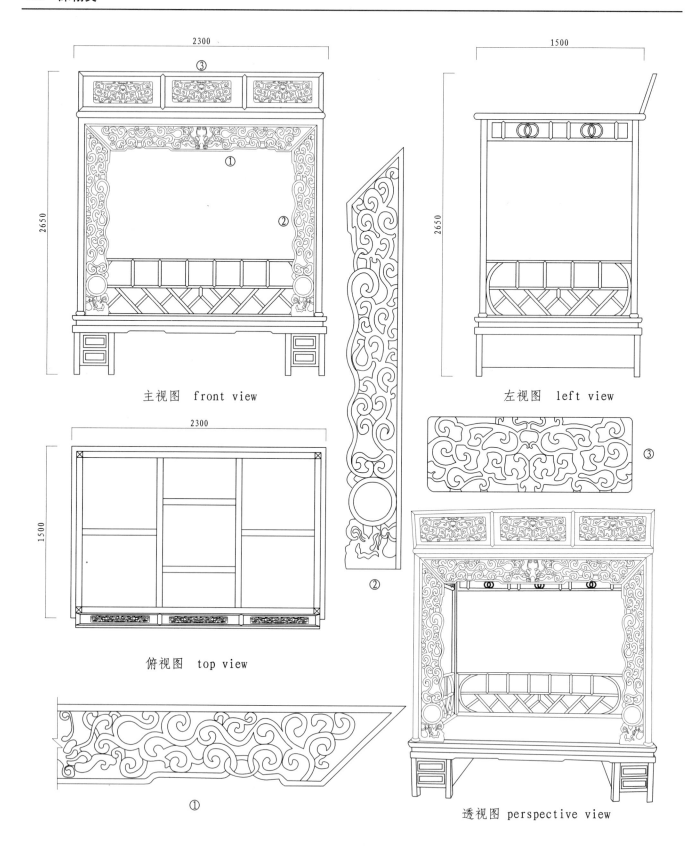

明代卷草纹大开门红木床
Ming dynasty *hong* wood canopy bed with wide opening and curling tendril design

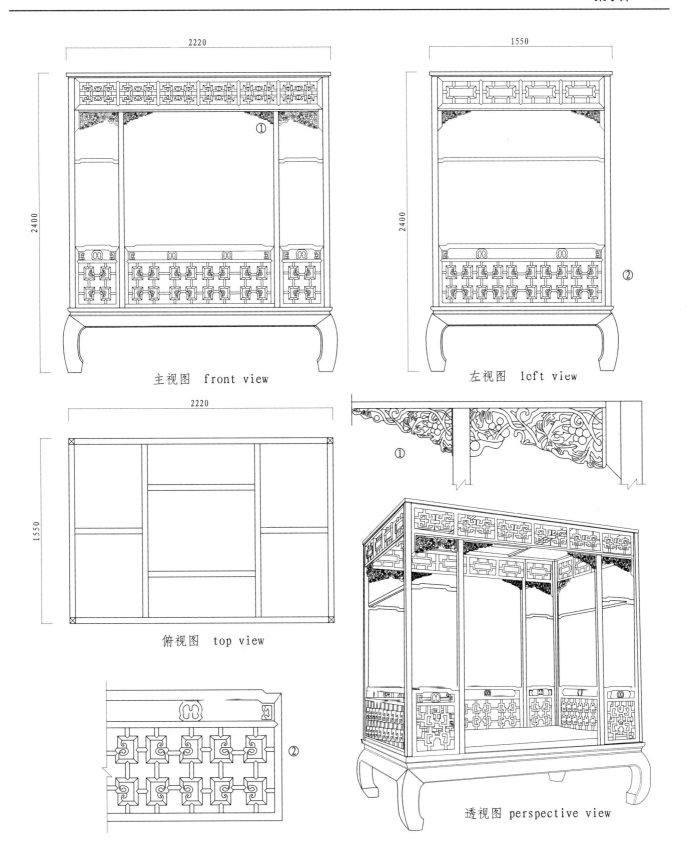

清乾隆红木架子床
Qing dynasty Qianlong period *hong* wood canopy bed

主视图 front view

左视图 left view

俯视图 top view

透视图 perspective view

明代黄花梨罗汉床
Ming dynasty *huanghuali* wood Luohan bed

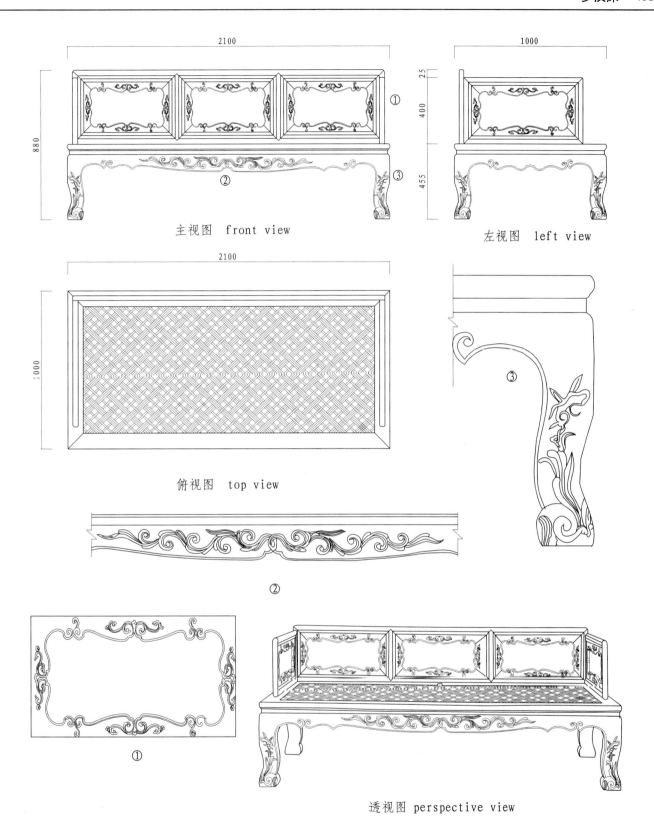

主视图 front view 左视图 left view

俯视图 top view

透视图 perspective view

明代黄花梨卷草纹藤心罗汉床
Ming dynasty *huanghuali* wood Luohan bed with curling tendril design and mat seat

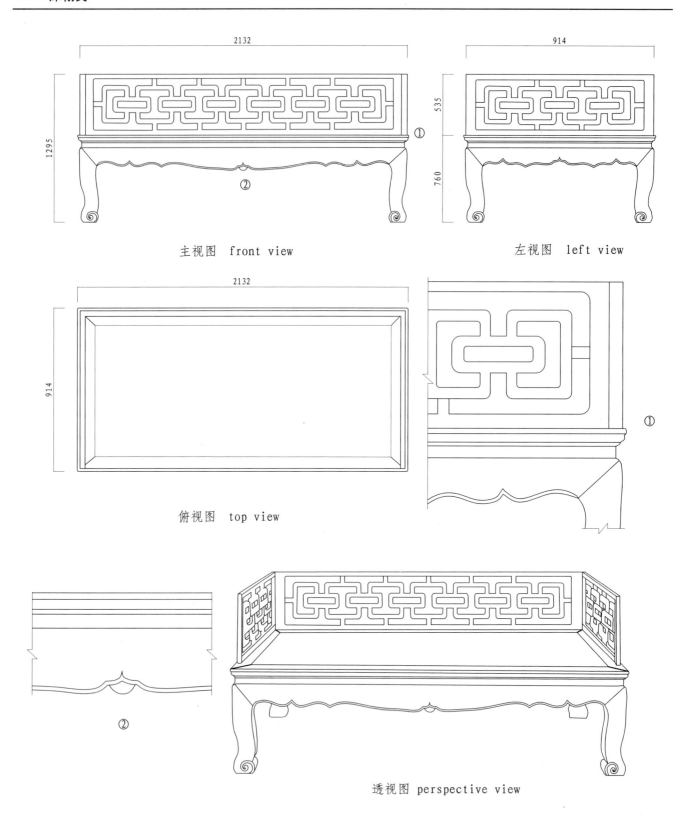

主视图 front view
左视图 left view
俯视图 top view
透视图 perspective view

清代黄花梨罗汉床
Qing dynasty *huanghuali* wood Luohan bed

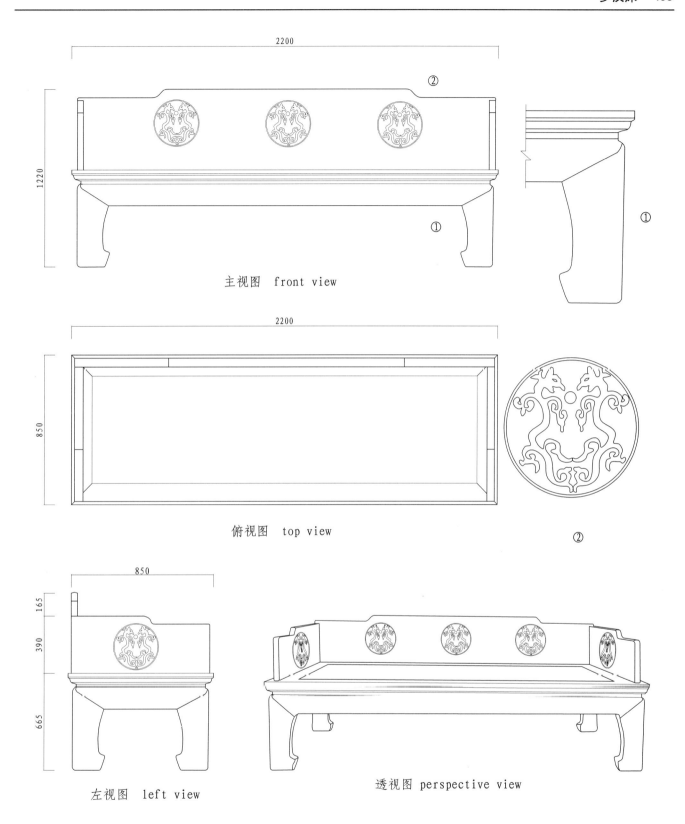

清中期榆木开光浮雕龙纹罗汉床
Mid Qing dynasty elm wood Luohan bed with medallion and relief carving of dragon design

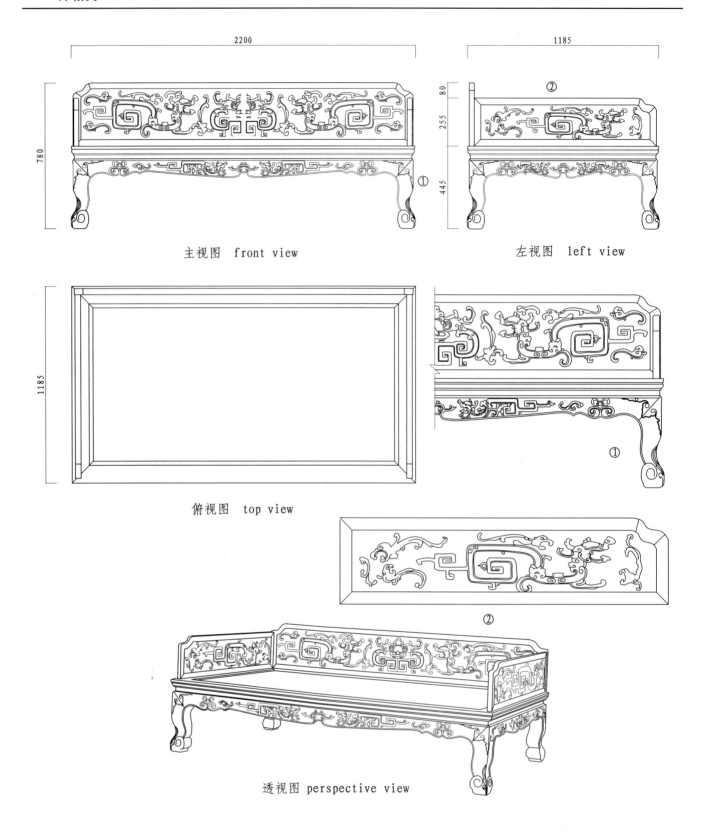

明代黄花梨三弯腿龙纹罗汉床
Ming dynasty *haunghuali* wood Luohan ned with cabriole legs and dragon design

主视图 front view

左视图 left view

俯视图 top view

透视图 perspective view

清代红木嵌大理石罗汉床
Qing dynasty *hong* wood Luohan bed with marble inlay

主视图 front view

俯视图 top view

左视图 left view

清雍正紫漆描金山水纹床
Qing dynasty Yongzheng period purple lacquer Luohan bed painting in gold with landscape design

透视图 perspective view

清雍正紫漆描金山水纹床
Qing dynasty Yongzheng period purple lacquer Luohan bde painting in gold with landscape design

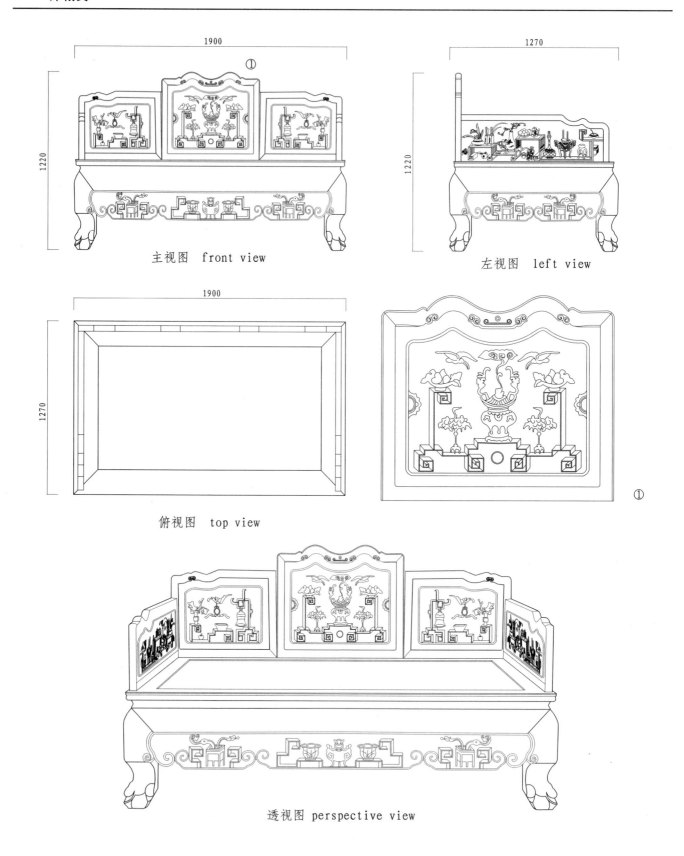

主视图 front view

左视图 left view

俯视图 top view

透视图 perspective view

清代酸枝木博古纹罗汉床
Qing dynasty *shuanzhi* wood waisted Luohan bed with five panel screens design of numerous antique precious objects

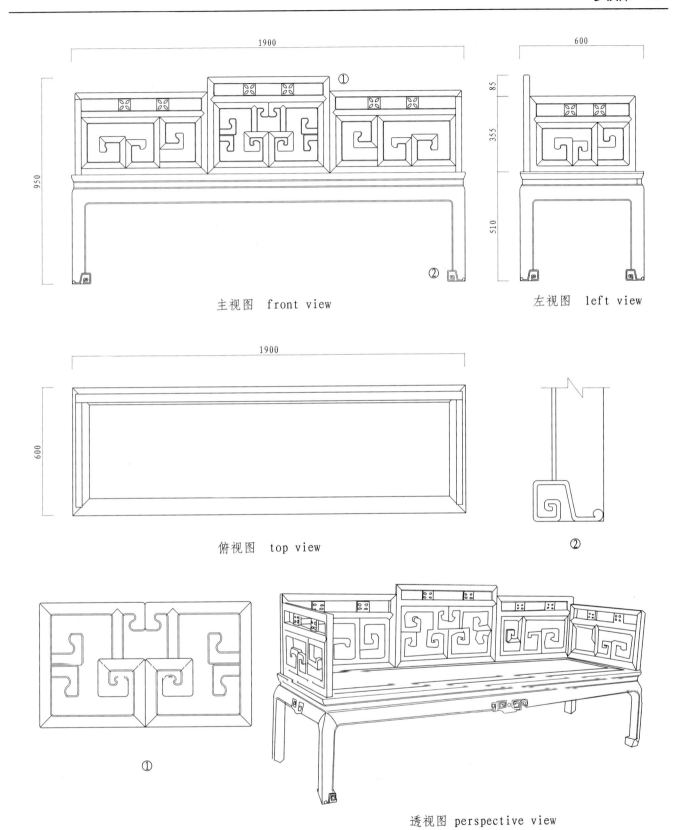

主视图 front view

左视图 left view

俯视图 top view

透视图 perspective view

清代酸枝三围屏卷云纹半床
Qing dynasty *suanzhi* wood half bed with three-panel screen railings and scrolled cloud motif

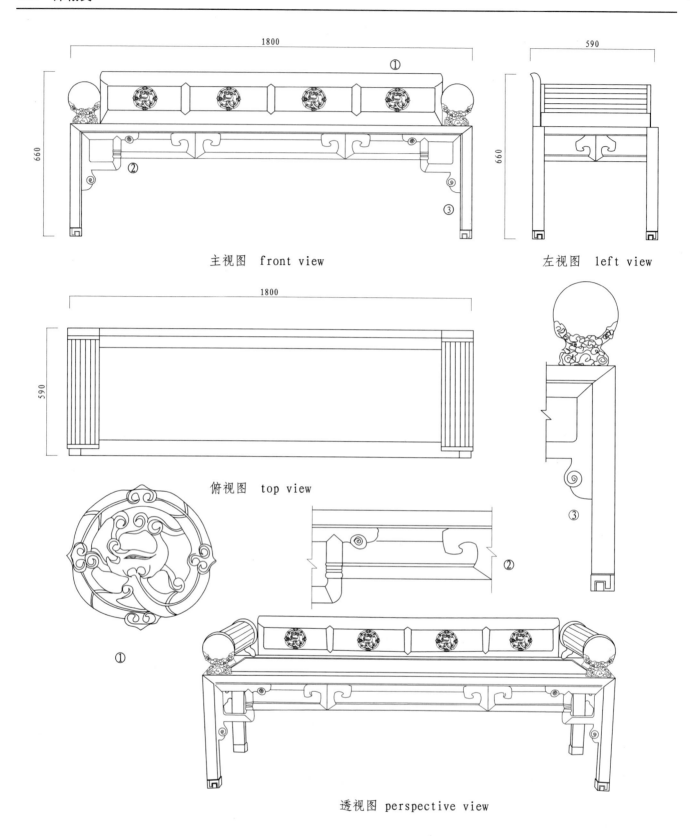

清代红木带枕凉床
Qing dynasty *hong* wood cool bed with pillows

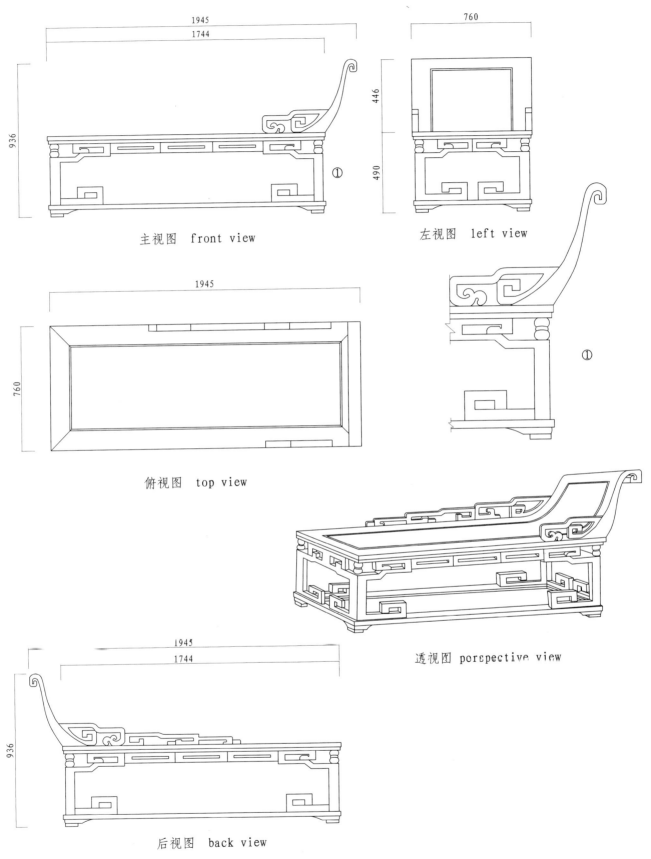

明代硬拐纹美人榻
Ming dynasty hard rectangular spiral pattern daybed for beautiful women

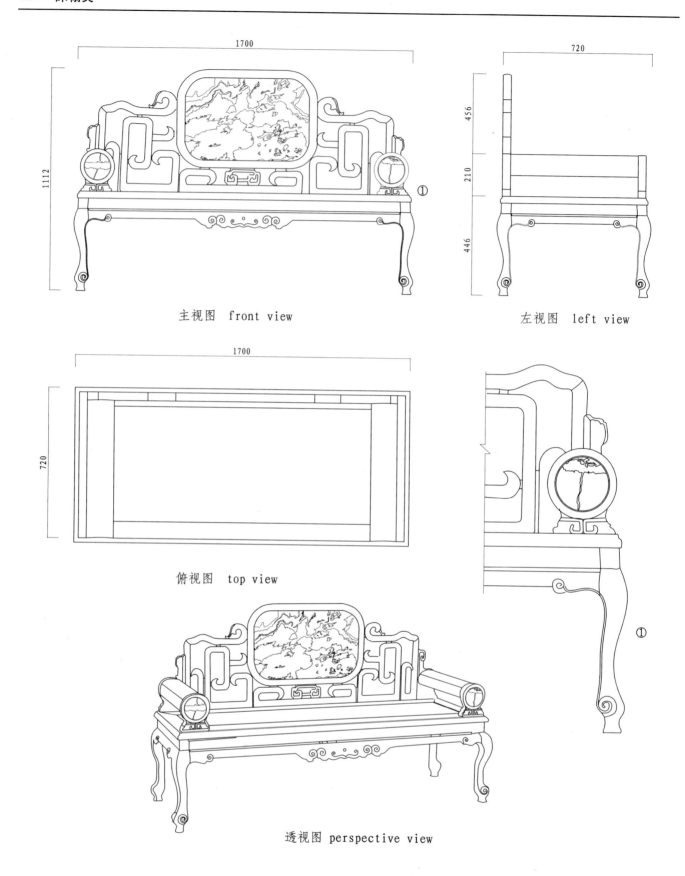

主视图 front view

左视图 left view

俯视图 top view

透视图 perspective view

清末年红木嵌大理石美人榻
Late Qing dynasty *hong* wood daybed inlaid with marble for beautiful women

衣柜 205

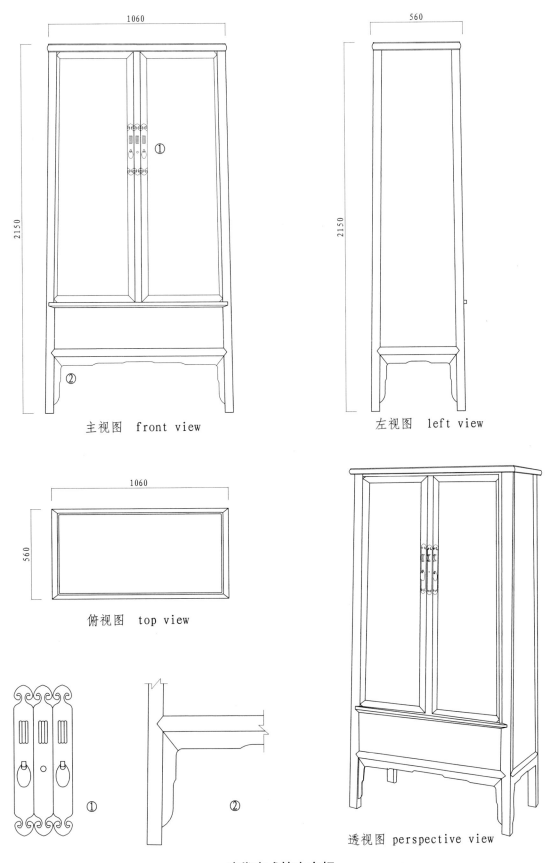

主视图 front view

左视图 left view

俯视图 top view

透视图 perspective view

清代宁式楠木衣柜
Qing dynasty Ningbo style *nan* wood wardrobe

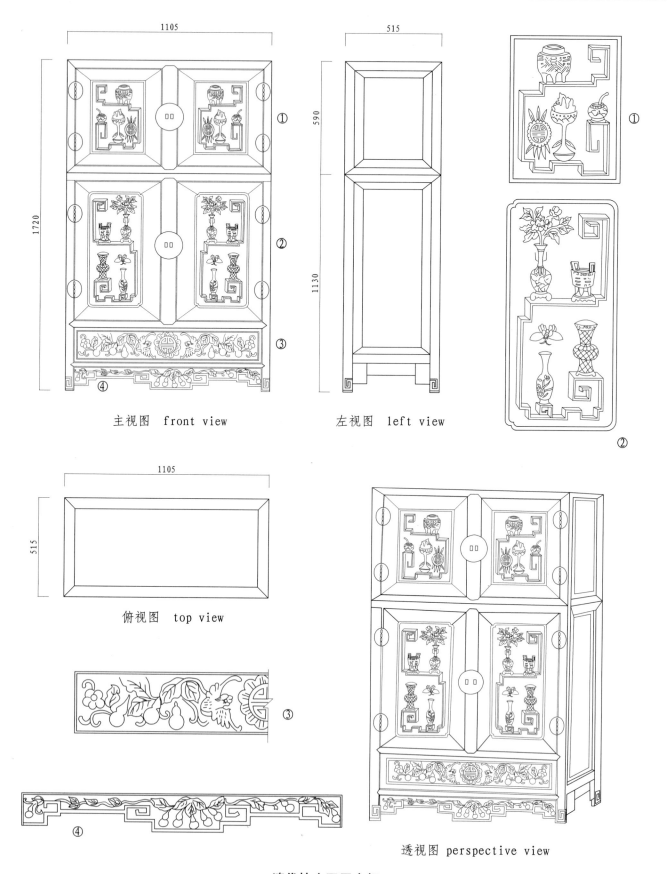

主视图 front view
左视图 left view
俯视图 top view
透视图 perspective view

清代楠木双层衣柜
Qing dynasty *nan* wood two-layer wardrobe

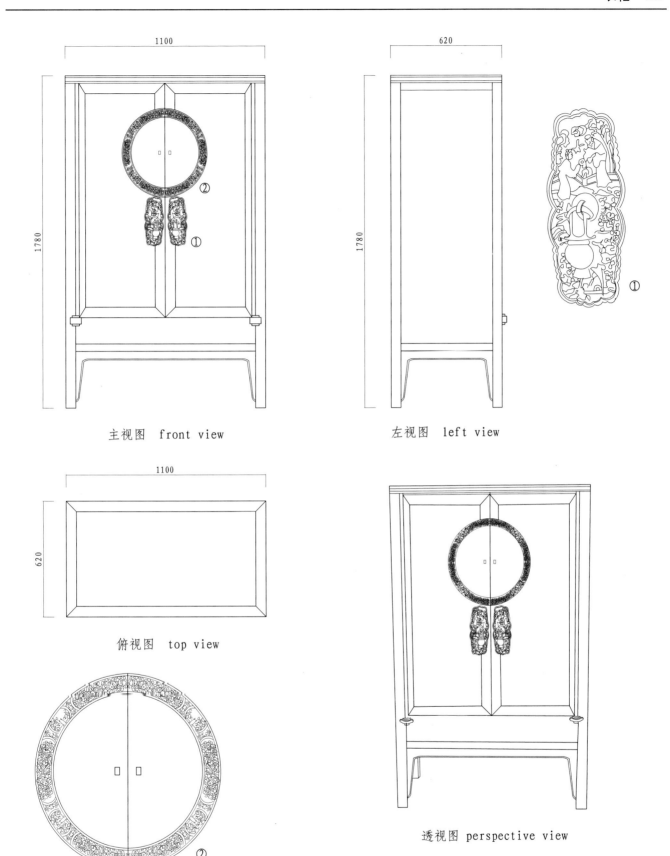

主视图 front view

左视图 left view

俯视图 top view

透视图 perspective view

清代宁式红衣橱
Qing dynasty Ningbo style red cabinet

主视图 front view　　左视图 left view

俯视图 top view

透视图 perspective view

明代黄花梨翘头草龙纹联二橱
Ming dynasty *huanghuali* wood two-drawer coffer with everted flanges and curling limbed dragon design

主视图 front view

左视图 left view

俯视图 top view

透视图 perspective view

明代黄花梨二屉联二橱
Ming dynasty *huanghuali* wood two-drawer coffer

主视图 front view　　左视图 left view

俯视图 top view

透视图 perspective view

明代翘头雕花联三柜
Ming dynasty three-drawer coffer with everted flanges and ornamental carving

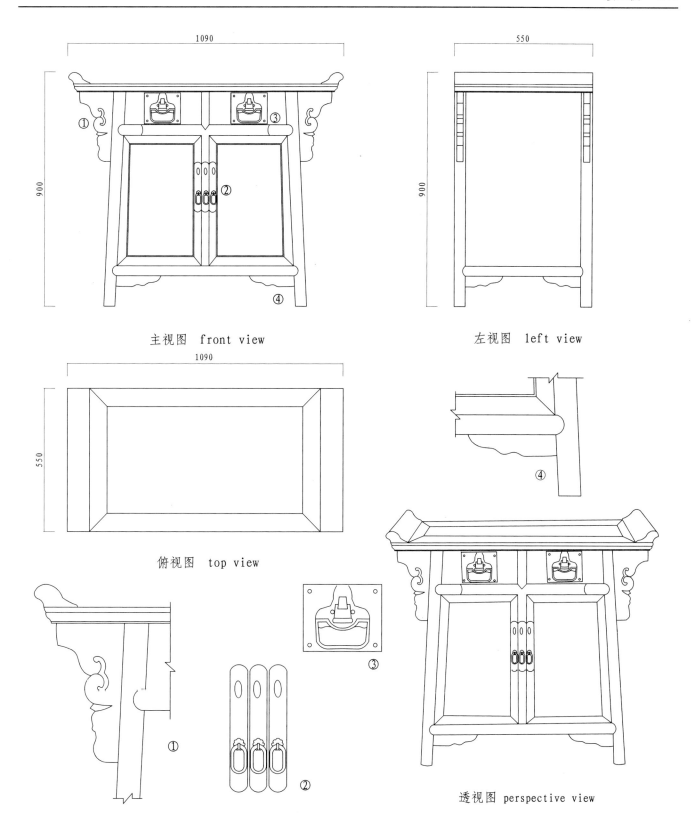

清代老花梨木老爷柜
Qing dynasty old *huali* wood two-drawer coffer

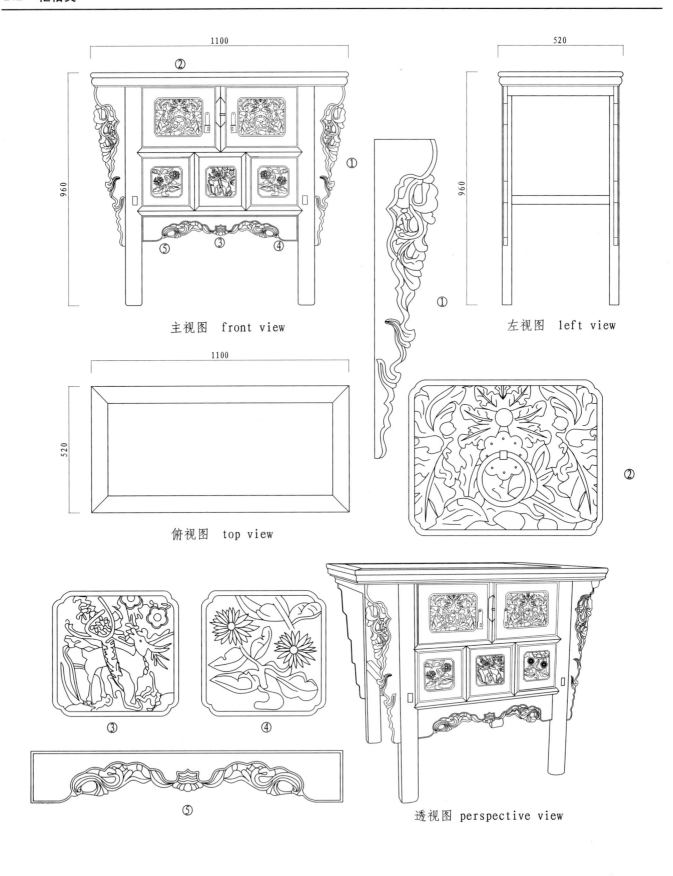

主视图 front view

左视图 left view

俯视图 top view

透视图 perspective view

清代榆木雕花闷橱

Qing dynasty elm wood two-drawer coffer with flower pattern carving

清代酸枝木夔龙纹五斗橱
Qing dynasty *suanzhi* wood cabinet with five drawers and kui-dragon design

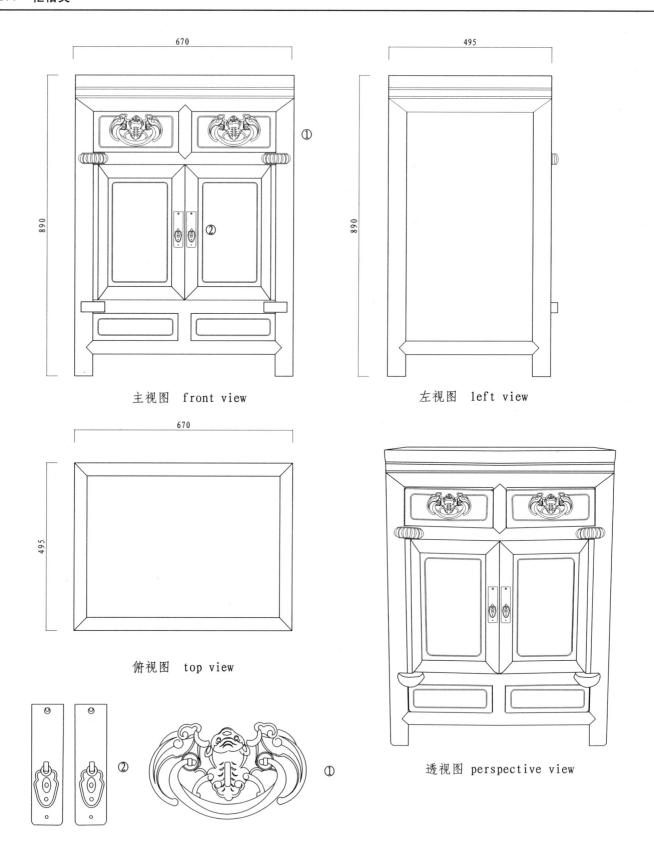

主视图 front view

左视图 left view

俯视图 top view

透视图 perspective view

清代宁式楠木床头柜
Qing dynasty Ningbo style *nan* wood bedstand

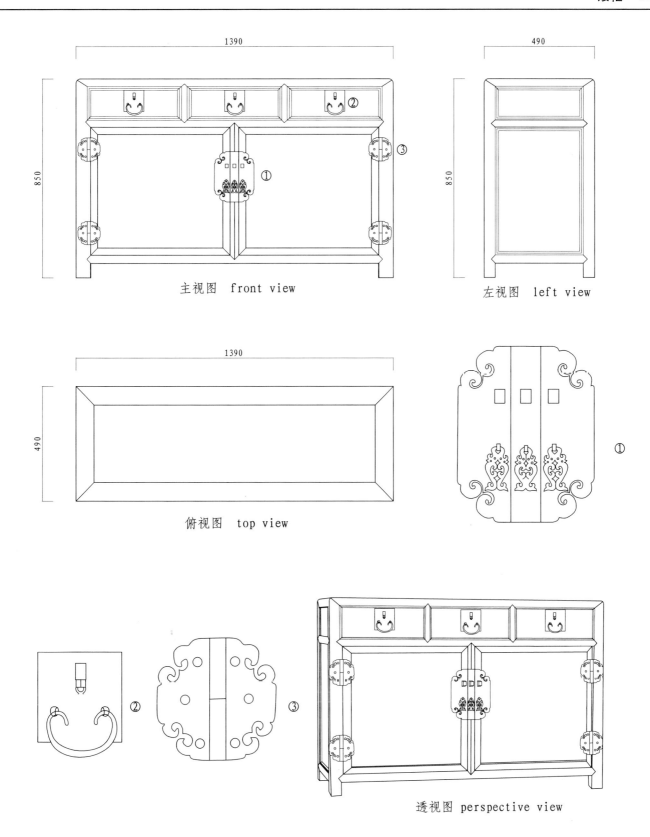

清代老花梨木矮柜
Qing dynasty old *huali* wood low cabinet

216 柜格类

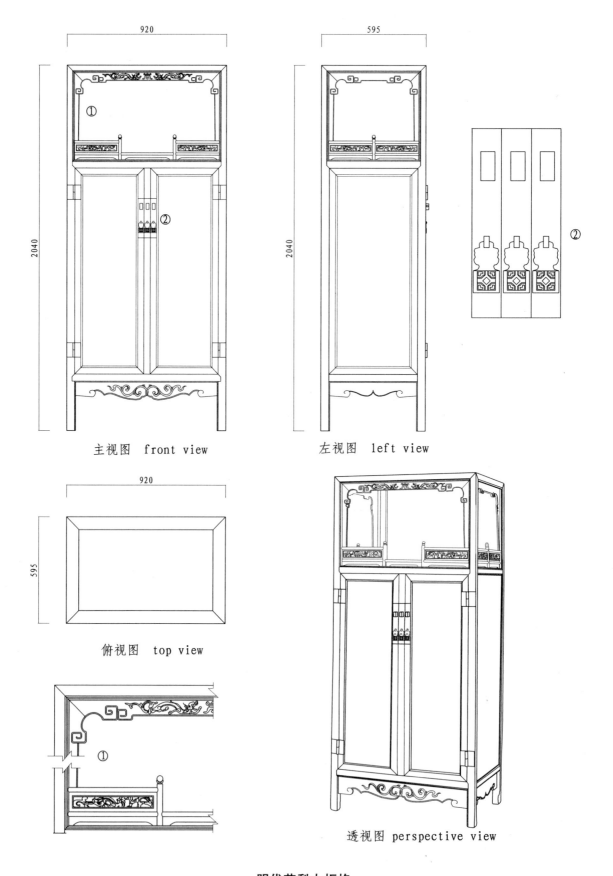

主视图 front view
左视图 left view
俯视图 top view
透视图 perspective view

明代花梨木柜格
Ming dynasty *huali* wood display cabinet

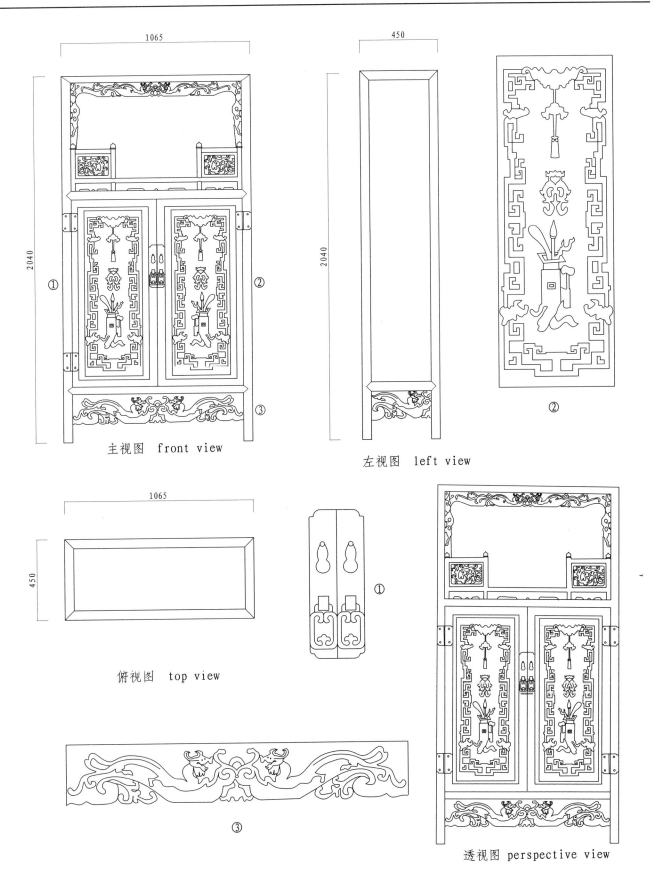

主视图 front view
左视图 left view
俯视图 top view
透视图 perspective view

清代黄花梨柜格
Qing dynasty haunghuali wood display cabinet

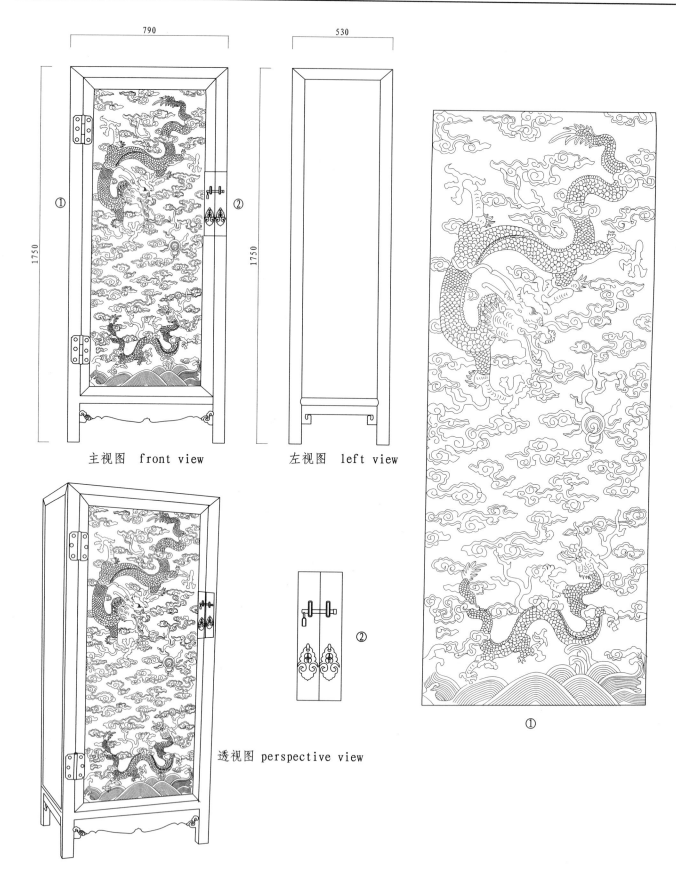

清初期黄花梨海水云龙纹单门柜
Early Qing dynasty *huanghuali* wood one-door cabinet with sea-and-dragon pattern design

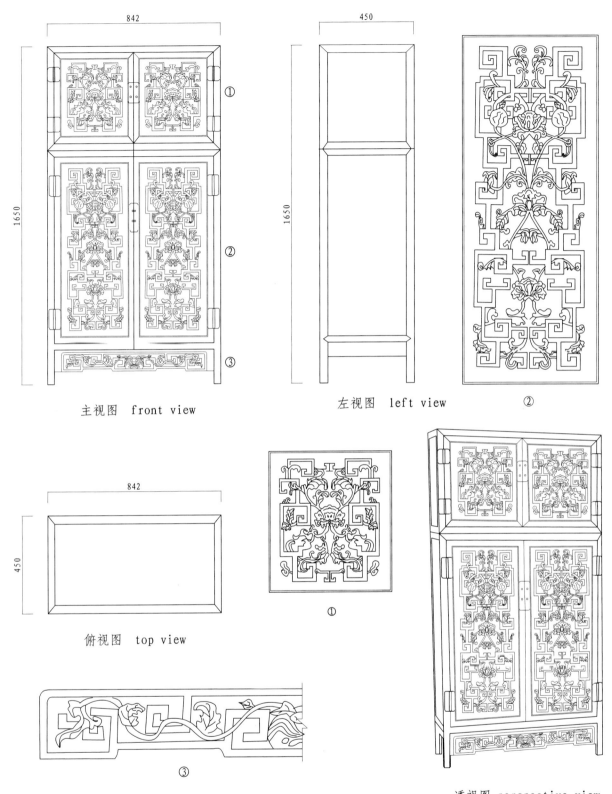

主视图 front view

左视图 left view

俯视图 top view

透视图 perspective view

清代黄花梨雕花顶竖柜
Qing dynasty *huanghuali* wood compound wardrobe in four parts with ornamental carving

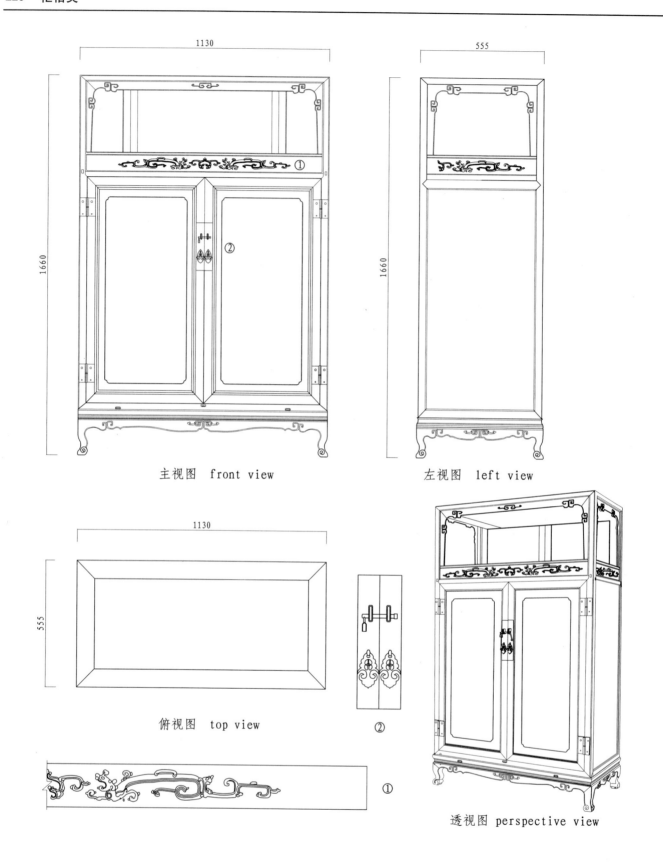

主视图 front view 左视图 left view 俯视图 top view 透视图 perspective view

明代黄花梨万历柜
Ming dynasty *huanghuali* wood Wanli display cabinet

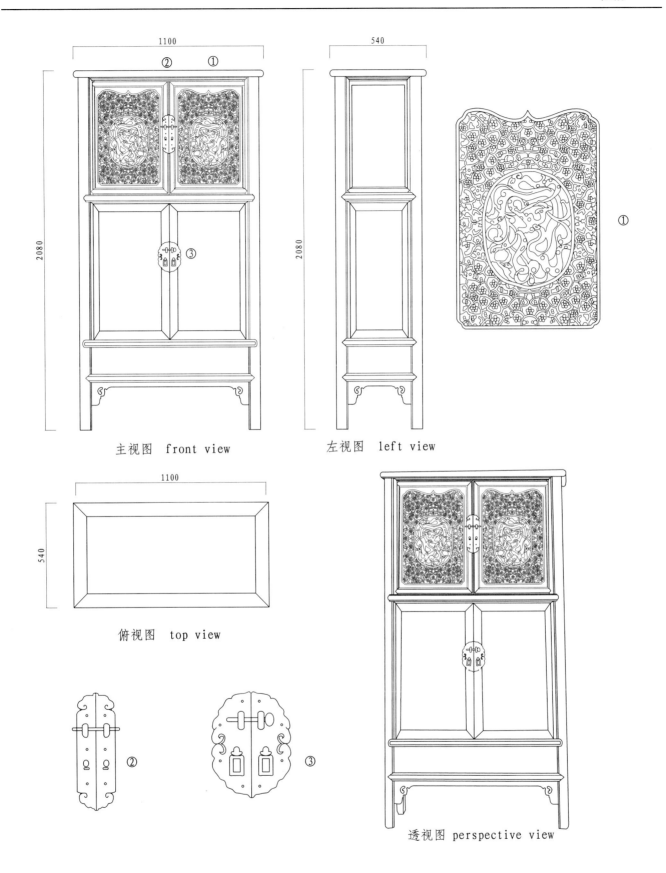

主视图 front view 左视图 left view

俯视图 top view

透视图 perspective view

清代榆木圆角柜
Qing dynasty elm wood round-corner cabinet

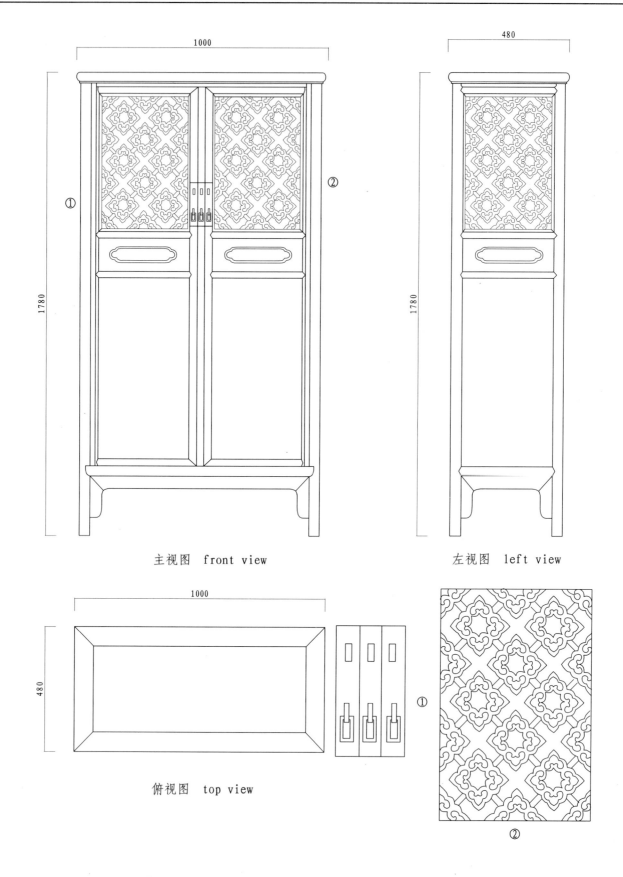

主视图 front view　　左视图 left view　　俯视图 top view

明代透格门圆角柜
Ming dynasty round-corner cabinet with lattice doors

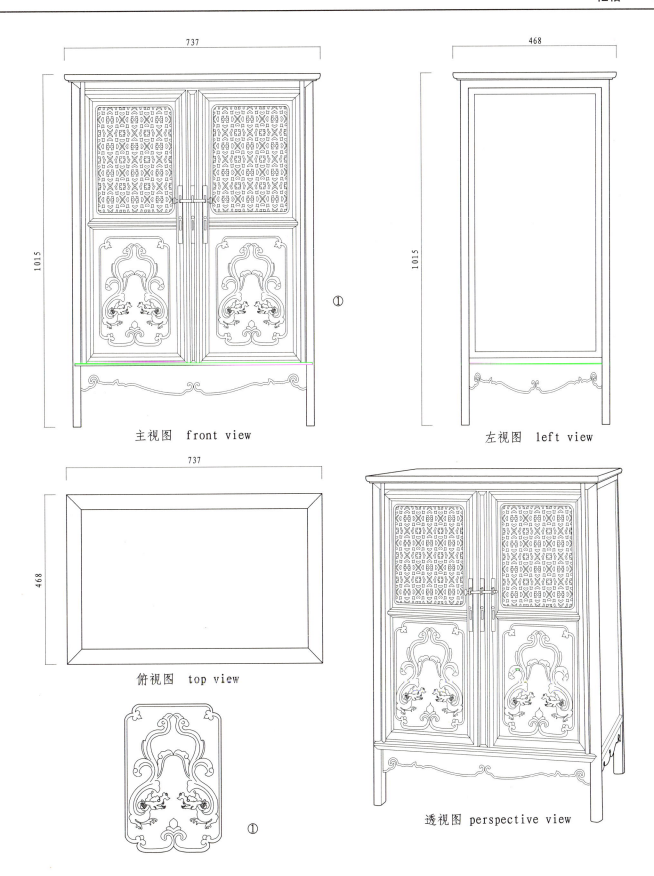

主视图 front view

左视图 left view

俯视图 top view

透视图 perspective view

清中期黄花梨圆角炕柜
Mid Qing dynasty *huanghuali* wood round-corner *kang* cabinet

主视图 front view　　左视图 left view

俯视图 top view

透视图 perspective view

清代核桃木人物二门中号柜
Qing dynasty walnut wood two-door medium size cabinet with figures design

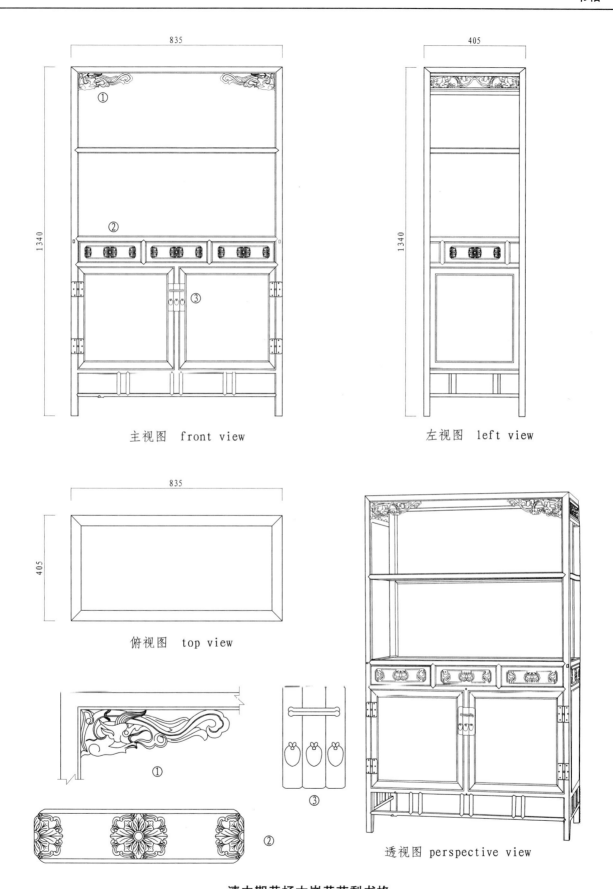

主视图 front view　左视图 left view　俯视图 top view　透视图 perspective view

清中期黄杨木嵌黄花梨书格
Mid Qing dynasty boxwood bookshelves inlaid with *huanghuali* wood

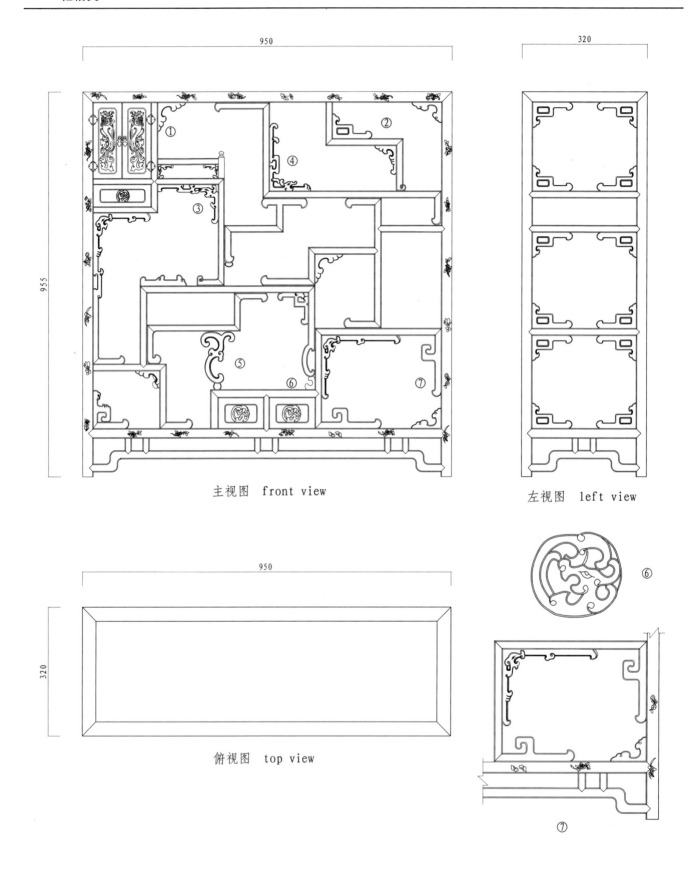

清乾隆楸木金夔凤纹多宝格
Qing dynasty Qianlong period *qiu* wood display cabinet with golden *kui*-phoenix motif

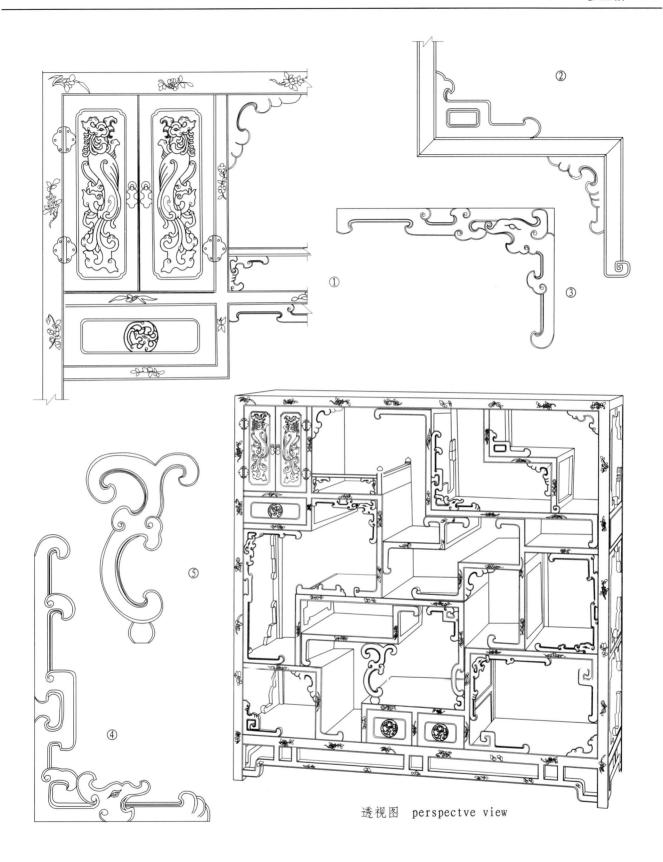

清乾隆楸木金夔凤纹多宝格
Qing dynasty Qianlong period *qiu* wood display cabinet with golden *kui*-phoenix motif

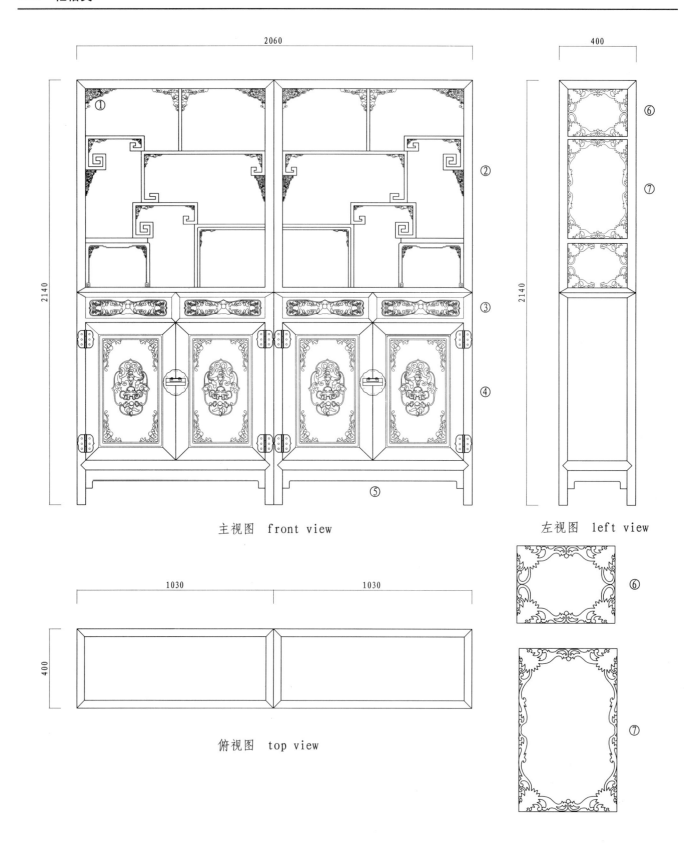

清代紫檀雕花多宝格
Qing dynasty *zitan* wood display cabinet with flower pattern carving

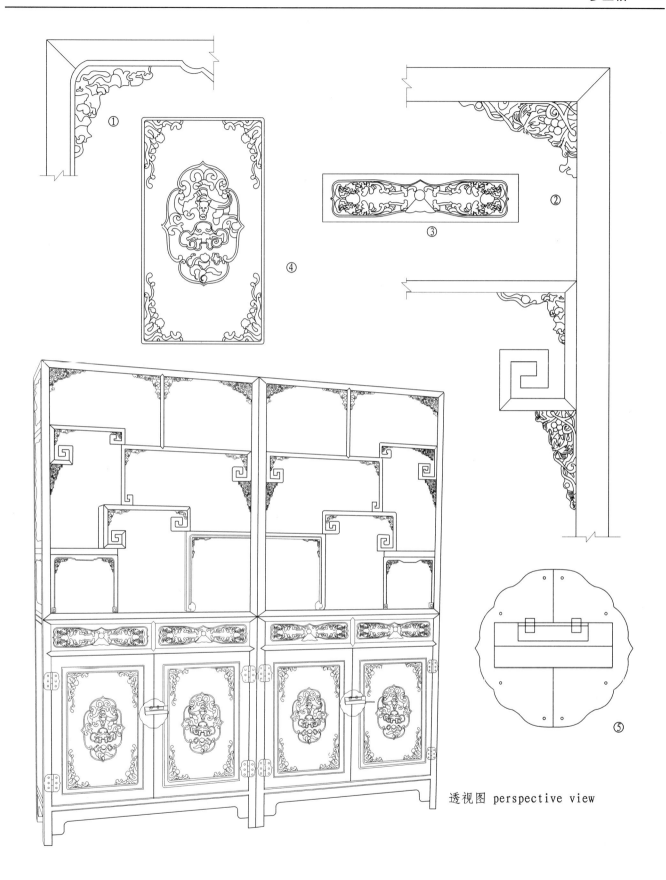

清代紫檀雕花多宝格
Qing dynasty *zitan* wood display cabinet with flower pattern carving

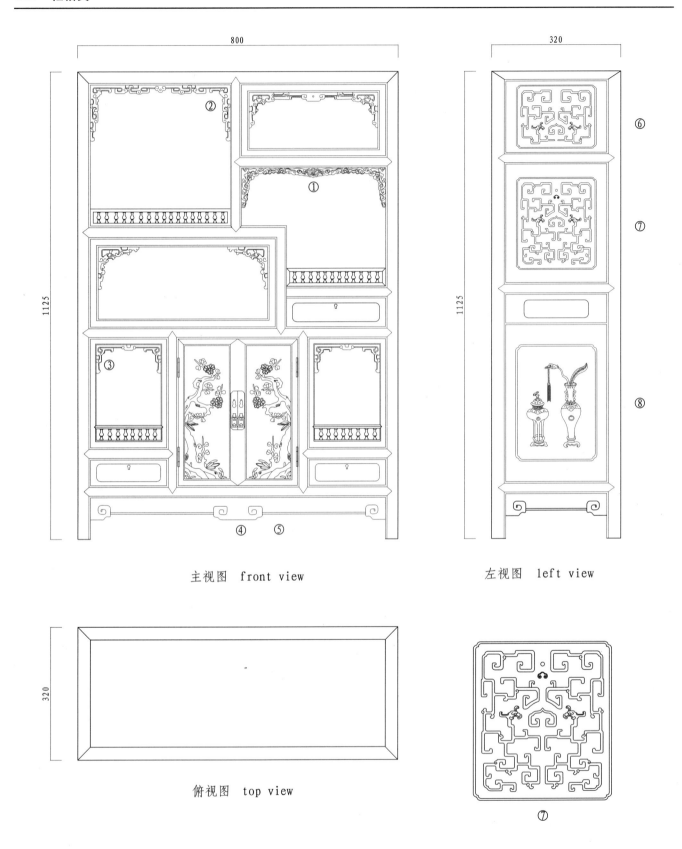

主视图 front view 左视图 left view 俯视图 top view

清代黄花梨梅花纹多宝格
Qing dynasty *huanghuali* wood display cabinet with plum blossom design

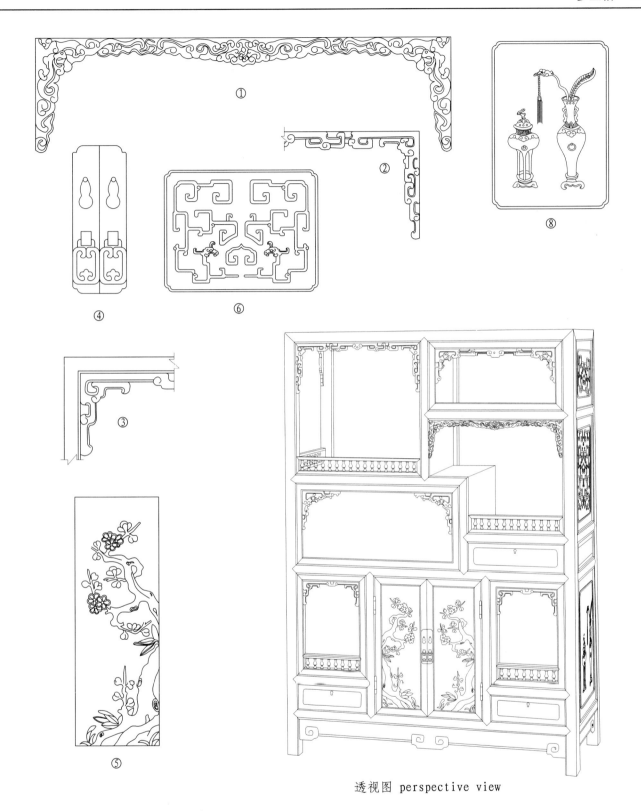

清代黄花梨梅花纹多宝格
Qing dynasty *huanghuali* wood display cabinet with plum blossom design

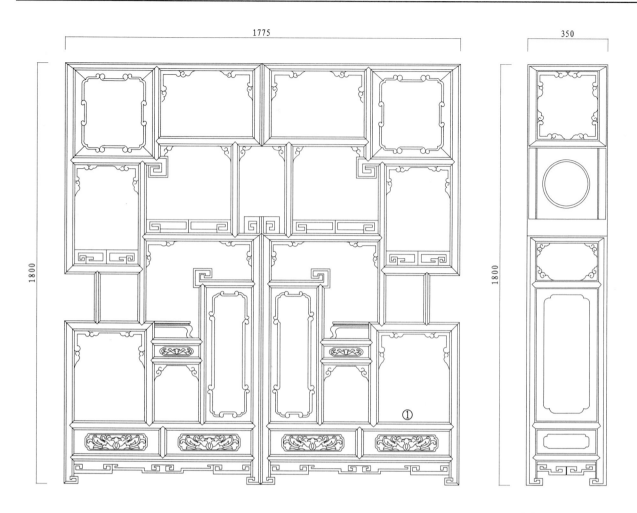

主视图 front view   左视图 left view

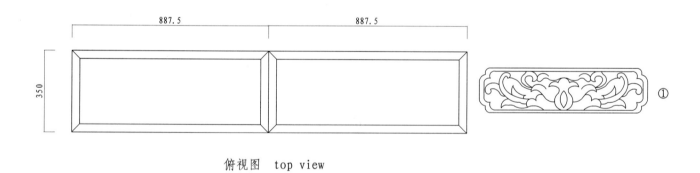

俯视图 top view

清代鸡翅木书柜式多宝格
Qing dynasty *jichi* wood display cabinet in the form of a book cabinet

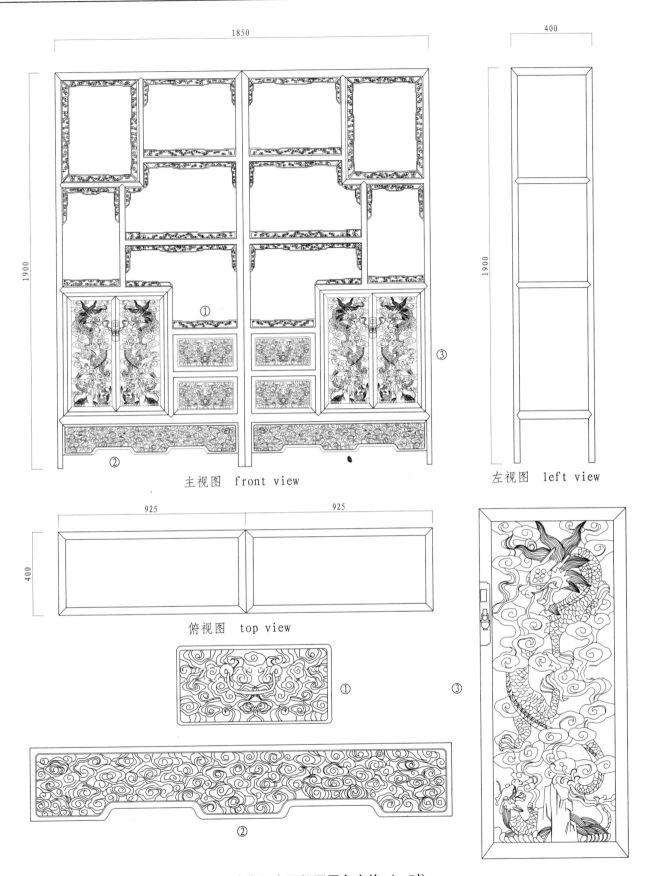

主视图 front view
左视图 left view
俯视图 top view

清代红木双门双屉多宝格（一对）
A pair of Qing dynasty *hong* wood display cabinet with double doors and drawers

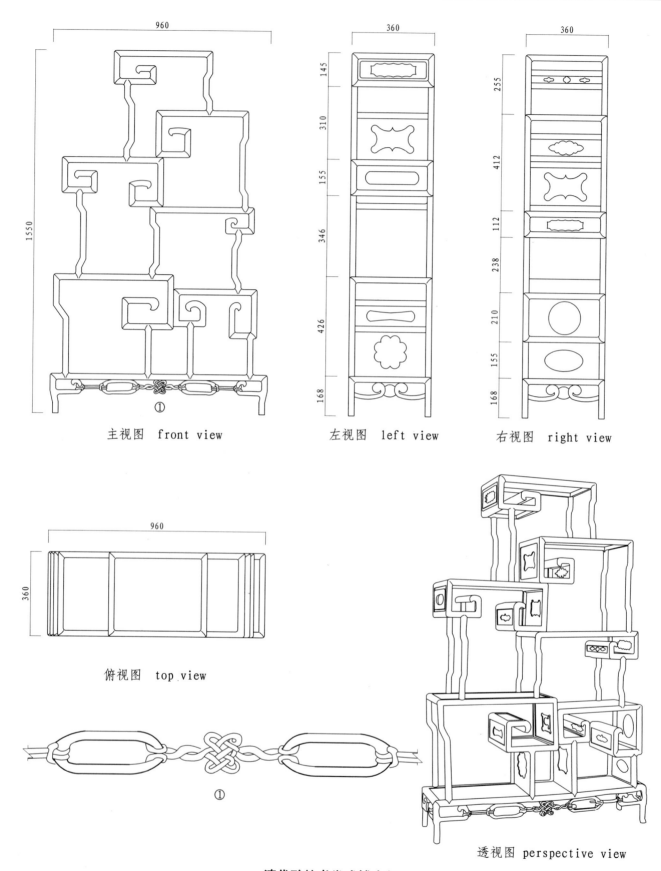

清代酸枝书卷式博古架
Qing dynasty *suanzhi* wood display shelves with cylindrical scroll

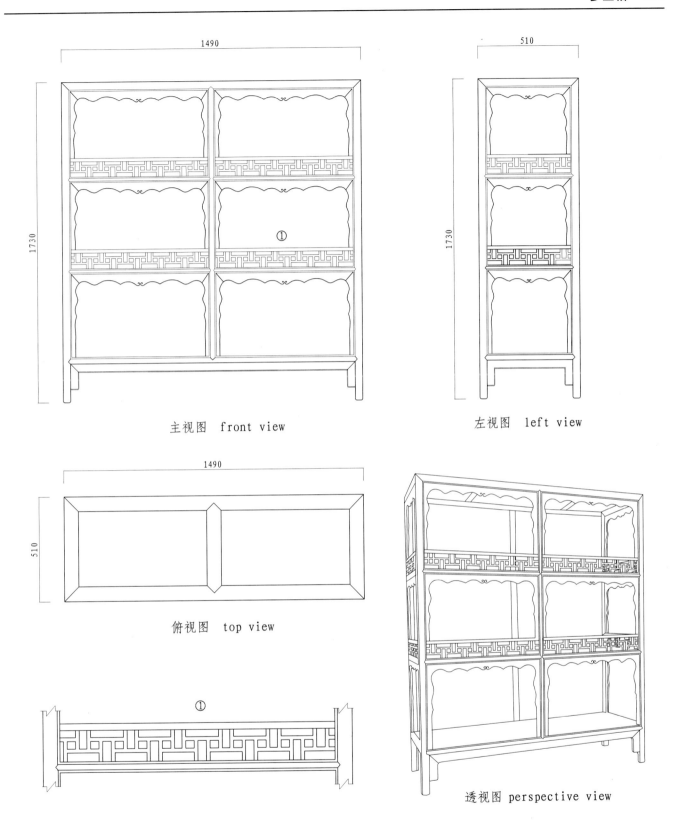

清代早期核桃木架格
Early Qing dynasty walnut wood shelves

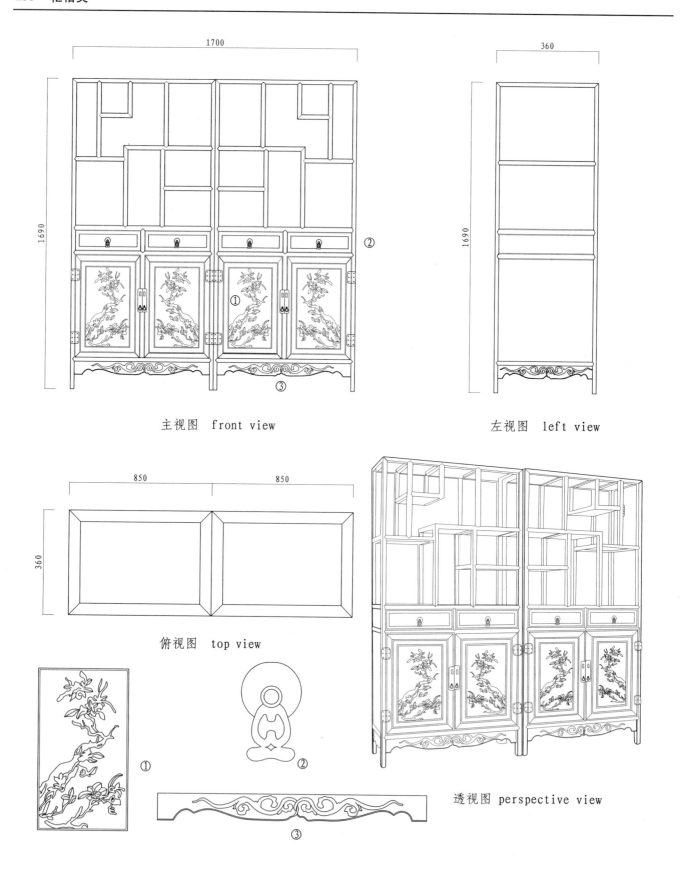

主视图 front view

左视图 left view

俯视图 top view

透视图 perspective view

清中期红木浮雕折枝花卉多宝格
Mid Qing dynasty *hong* wood display cabinet with relief carving of flower pattern

明代红漆雕鳞凤纹插屏
Ming dynasty red lacquer screen set in a stand with removable panel and scale-phoenix design carving

主视图 front view

左视图 left view

透视图 perspective view

清代紫檀百宝嵌插屏
Qing dynasty zitan wood screen set in a stand with removable panel and one-hundred-precious-material inlay

屏类 239

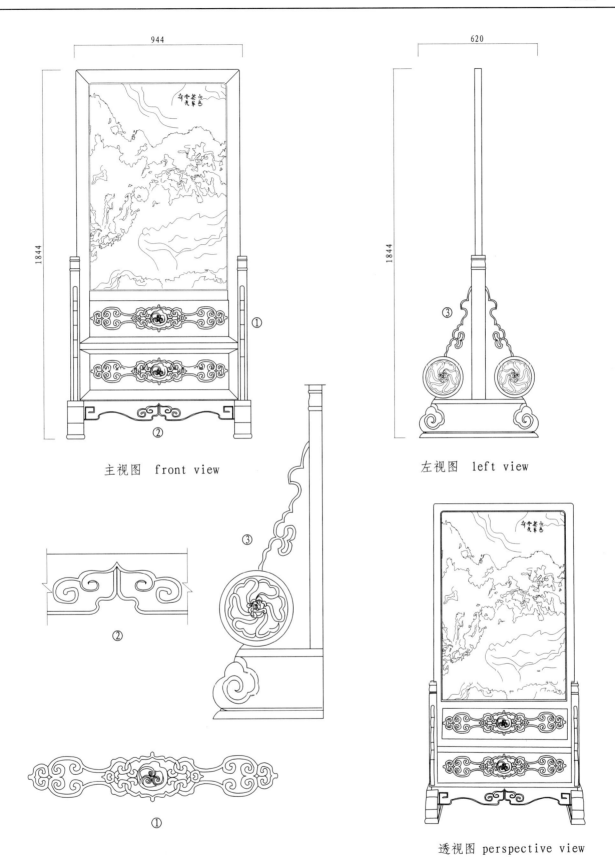

主视图 front view

左视图 left view

透视图 perspective view

清代黄花梨如意纹大理石插屏
Qing dynasty *huanghuali* wood screen set in a stand with removable marble panel and *ruyi* motif

清代红木拐子龙纹插屏

Qing dynasty *hong* wood screen set in a stand with removable panel and rectangular spiral dragon design

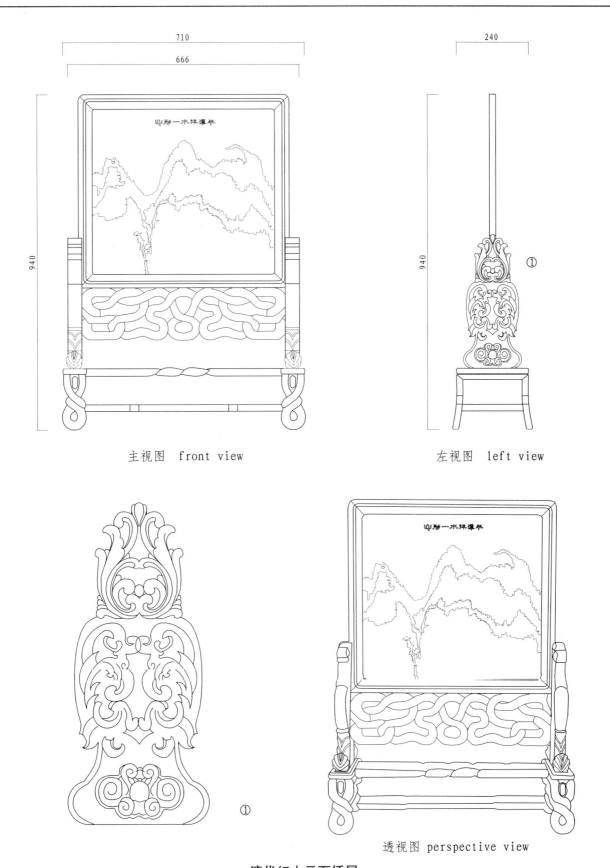

主视图 front view

左视图 left view

透视图 perspective view

清代红木云石插屏
Qing dynasty *hong* wood screen set in a stand with removable cloud design marble panel

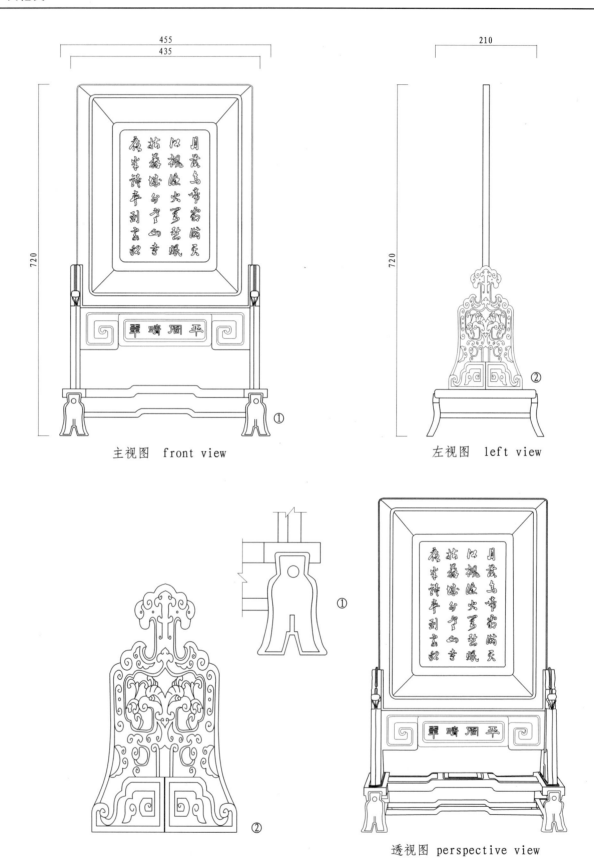

主视图 front view　　左视图 left view

透视图 perspective view

清代镶瘿木黄杨唐诗插屏
Qing dynasty boxwood screen set in a stand with burl wood inlay and removable panel of Tang dynasty poem design

清代花梨嵌玉璧插屏
Qing dynasty *huali* wood screen set in a stand with removable panel and round flat pieces of jade design inlay

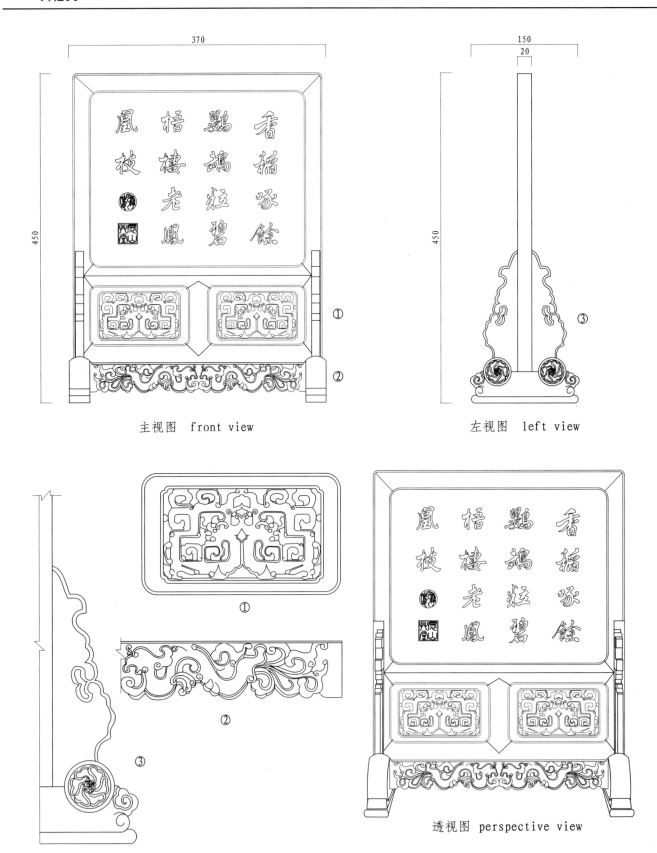

清代红木螺钿三狮进宝图插屏

Qing dynasty hong wood screen set in a stand with removable panel, mother-of-pearl inlay and three-lion-dedicating-treasure motif

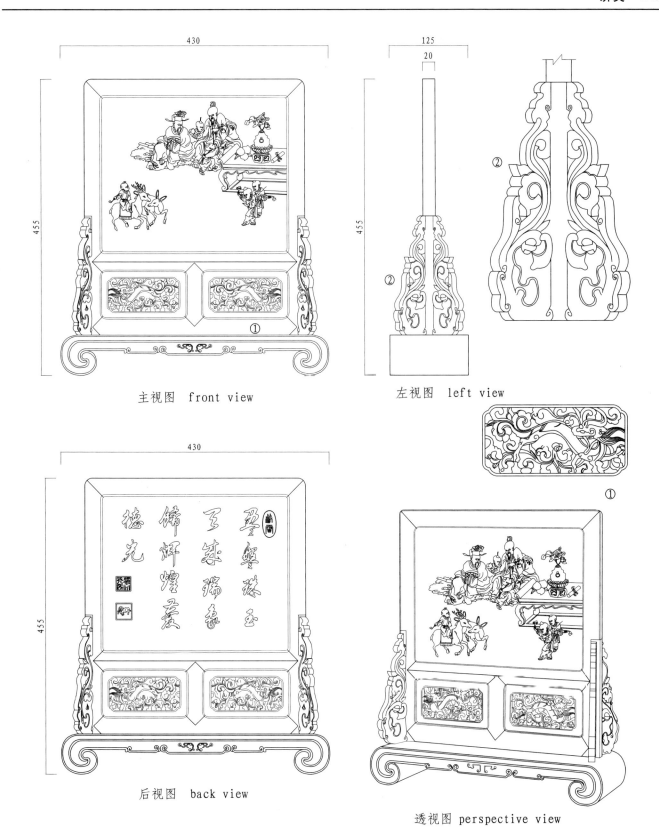

主视图 front view

左视图 left view

后视图 back view

透视图 perspective view

清代红木嵌螺钿三星图插屏
Qing dynasty *hong* wood screen set in a stand with removable panel, mother-of-pearl inlay and three immortals motif

清代紫檀夔龙纹插屏

Qing dynasty *zitan* wood screen set in a stand with removable panel and *kui*-dragon design

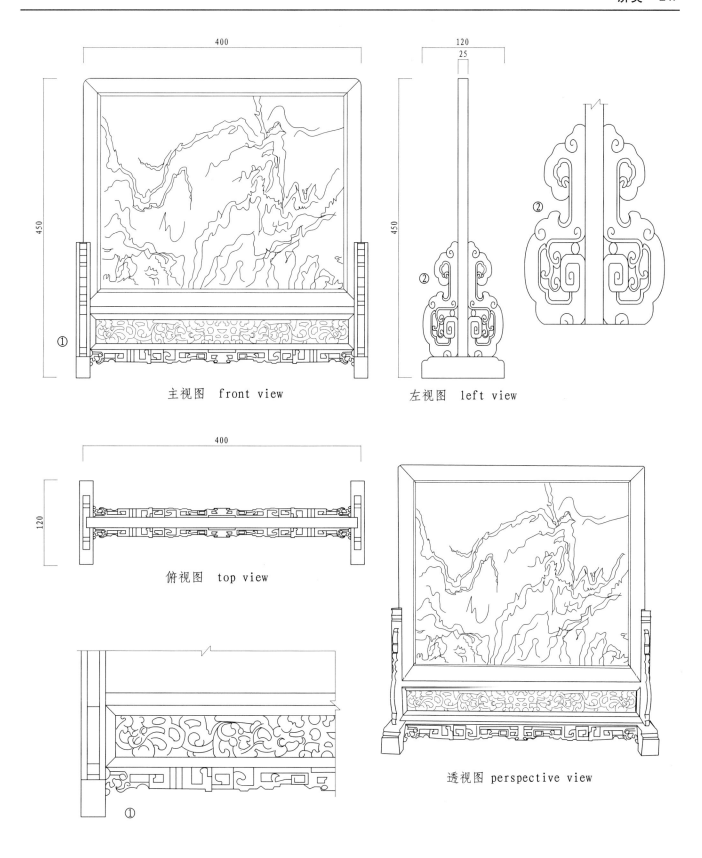

清代紫檀镶大理石坐屏
Qing dynasty zitan wood screen set in a stand with marble inlay

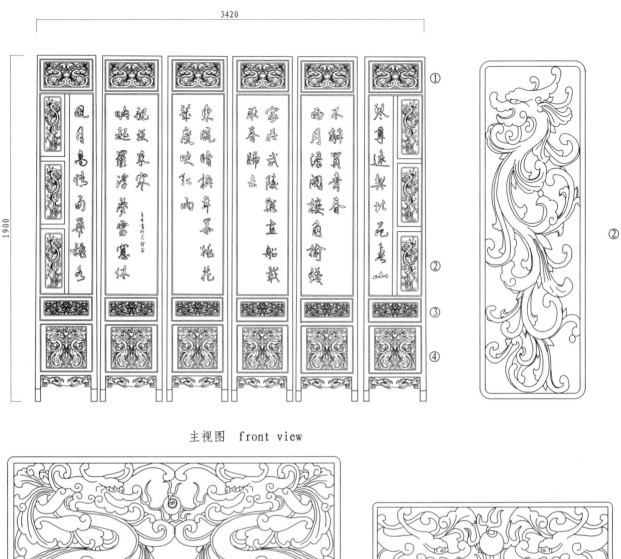

主视图　front view

清早期黄花梨透雕龙纹隔屏
Early Qing dynasty *huanghuali* wood screen with dragon design openwork carving

铁画四季花卉图挂屏。挂屏四扇成堂，硬木边框，屏心镶铁制牡丹、荷花、菊花、梅花等四季花卉，末署"汤鹏"二字并二方铁制印章款。

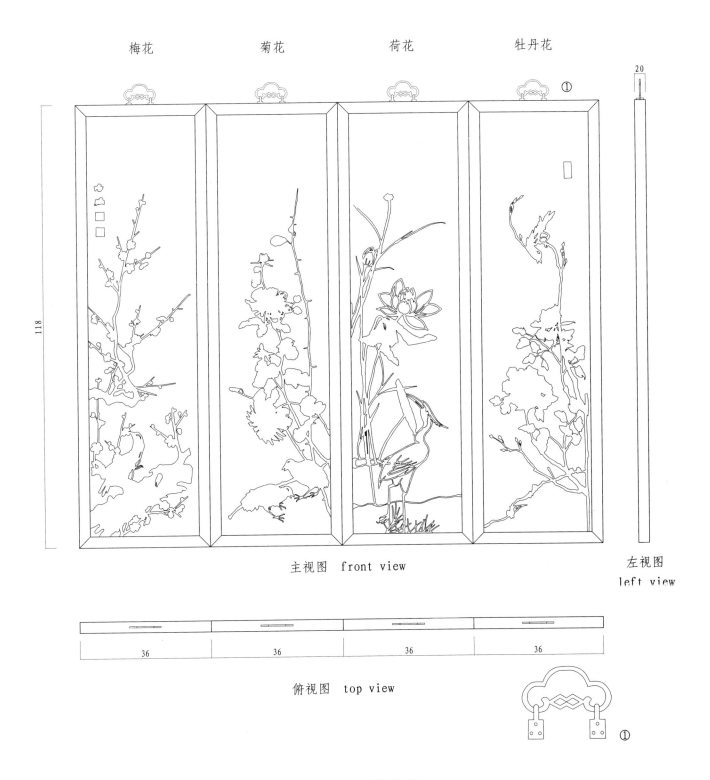

清康熙硬木铁画挂屏
Qing dynasty Kangxi period hard wood hanging screen with iron pictures

主视图 front view

透视图 perspective view

明代黄花梨浮雕花卉屏风
Ming dynasty *huanghuali* wood screen with relief carving of flower pattern

明代黄花梨雕双螭纹方台
Ming dynasty *huanghuali* wood square stand with double stylized hornless dragon design carving

主视图 front view

左视图 left view

俯视图 top view

清早期紫檀折叠镜台
Early Qing dynasty *zitan* wood folding mirror platform

透视图 perspective view

清早期紫檀折叠镜台

Early Qing dynasty *zitan* wood folding mirror platform

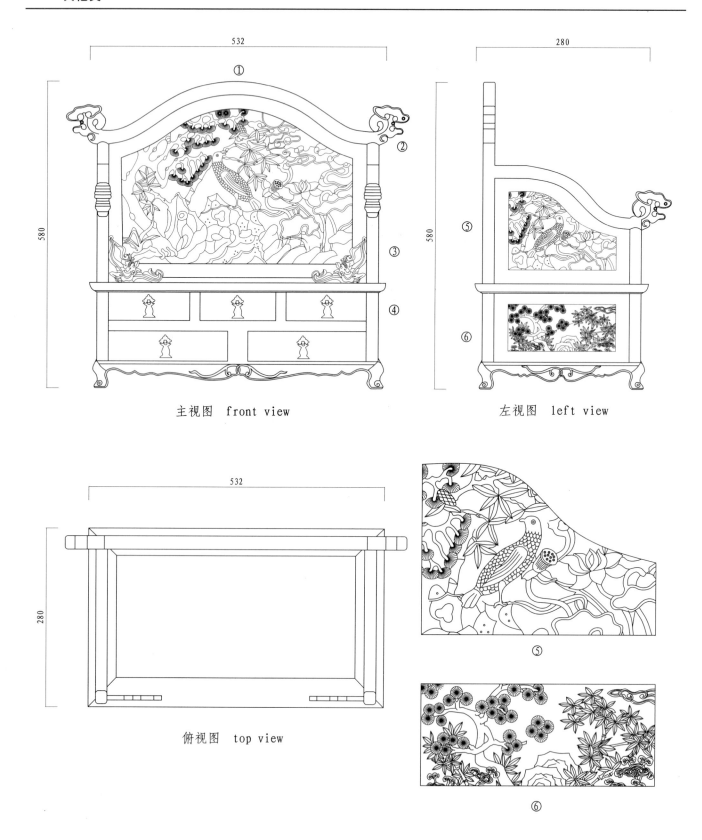

主视图 front view

左视图 left view

俯视图 top view

清代紫檀镂雕一品清廉纹镜台
Qing dynasty *zitan* wood mirror platform with openwork carving of *yipin-qinglian* motif

透视图 perspective view

清代紫檀镂雕一品清廉纹镜台
Qing dynasty *zitan* wood mirror platform with openwork carving of *yipin-qinglian* motif

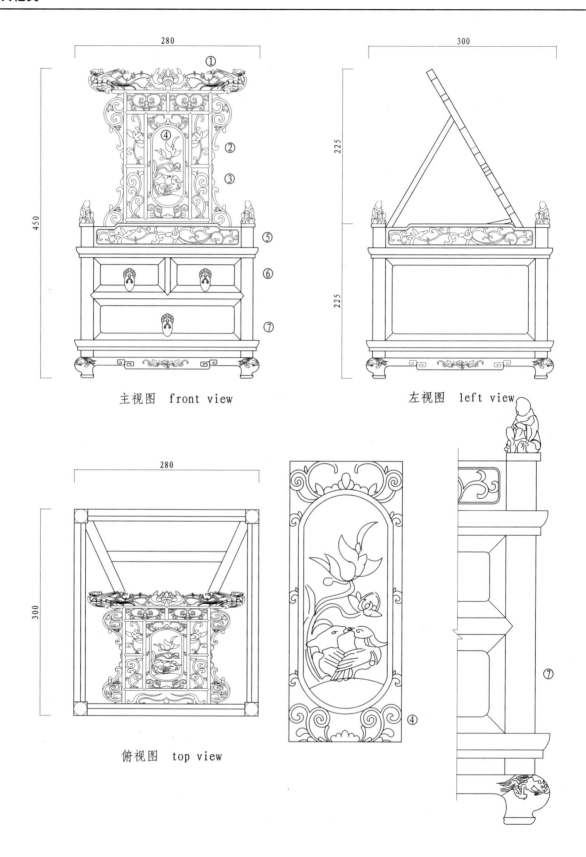

明代宁式黄花梨镜台
Ming dynasty Ningbo style *huanghuali* wood mirror platform

透视图 perspective view

明代宁式黄花梨镜台
Ming dynasty Ningbo style *huanghuali* wood mirror platform

主视图 front view

左视图 left view

俯视图 top view

明代黄花梨宝座式镂雕龙纹镜台
Ming dynasty *huanghuali* wood throne-type mirror platform with openwork carving of dragon design

透视图 perspective view

明代黄花梨宝座式镂雕龙纹镜台
Ming dynasty *huanghuali* wood throne-type mirror platform with openwork carving of dragon design

260 其他类

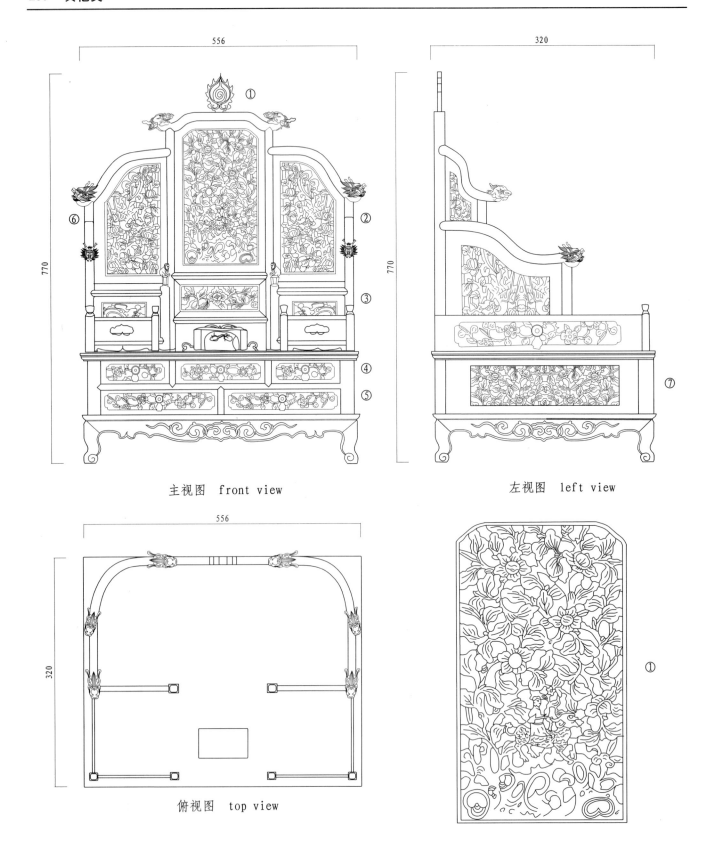

主视图 front view

左视图 left view

俯视图 top view

清代黄花梨西番莲花卉、麟凤纹五屏式镜台
Qing dynasty *huanghuali* wood mirror platform with five-panel screen
and craving design of passion flowers and scale-phoenix pattern

透视图 perspective view

清代黄花梨西番莲花卉、麟凤纹五屏式镜台
Qing dynasty *huanghuali* wood mirror platform with five-panel screen and craving design of passion flowers and scale-phoenix pattern

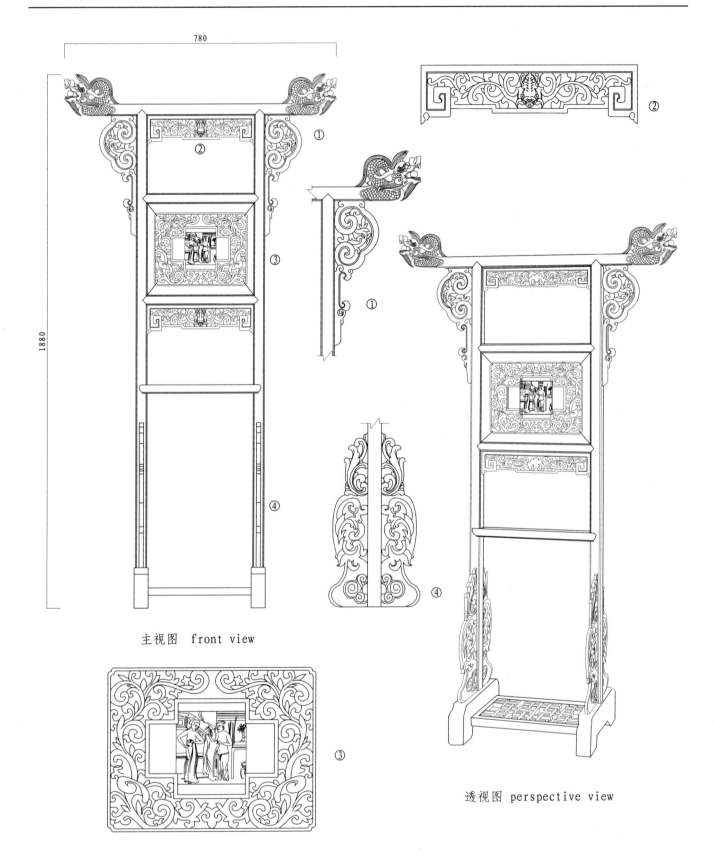

清代宁式楠木龙门衣架
Qing dynasty Ningbo style *nan* wood clothes racks with dragon designs

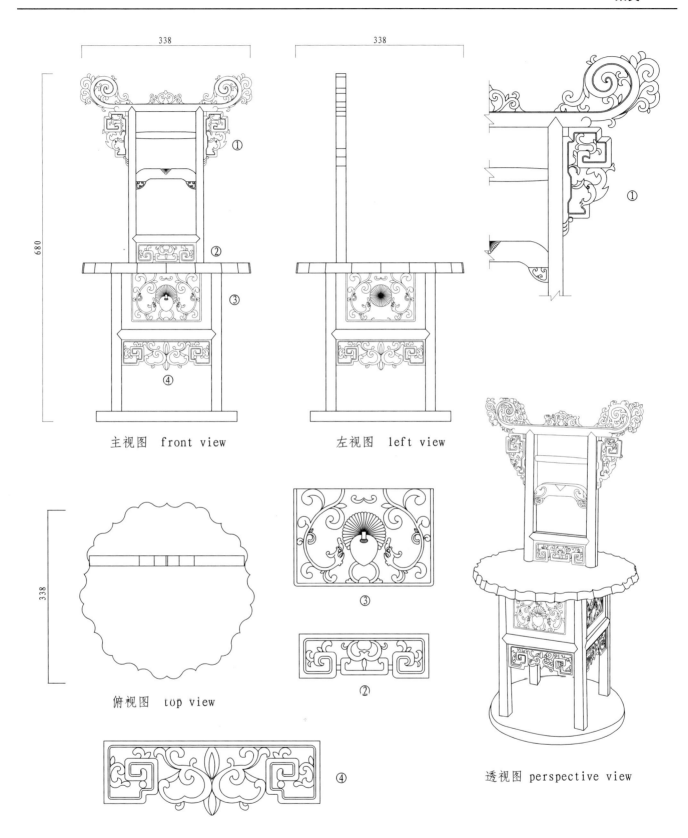

清代宁式榉木缠脚架
Qing dynasty Ningbo style *ju* wood stand for binding feet

264　其他类

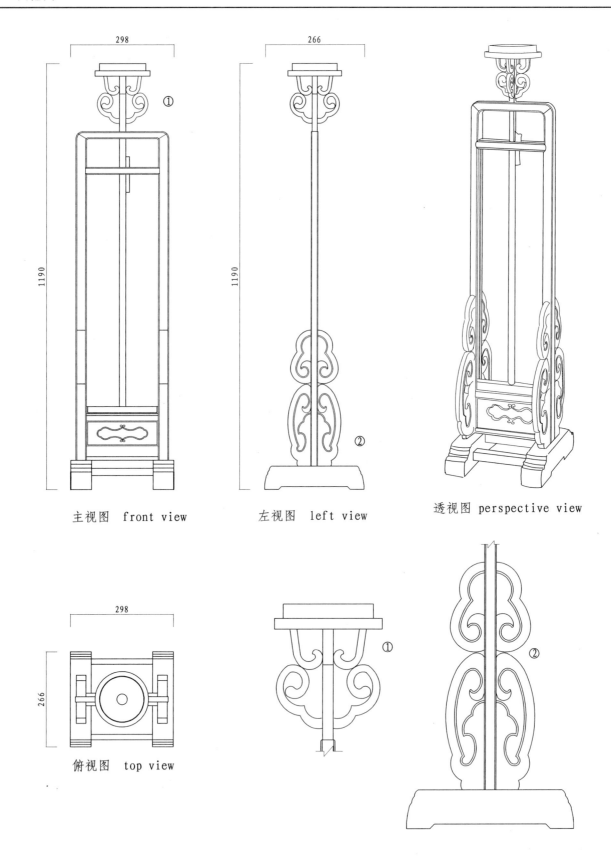

清代黄花梨木灯架
Qing dynasty *huanghuali* wood lamp stand

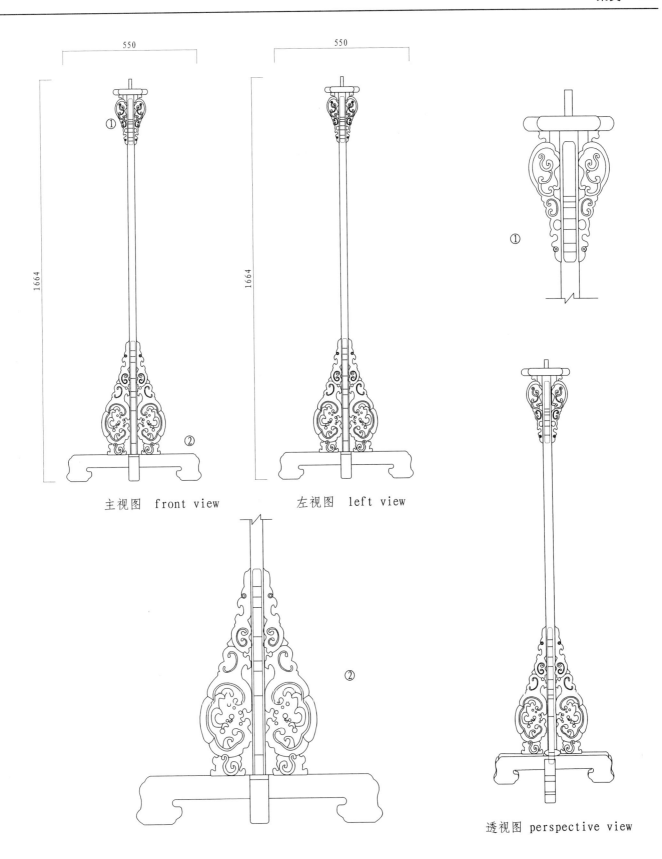

主视图 front view　　左视图 left view　　透视图 perspective view

清代紫檀灯架
Qing dynasty *zitan* wood lamp stand

主视图 front view

左视图 left view

俯视图 top view

透视图 perspective view

清代宁式柏木火盆架
Qing dynasty Ningbo style cypress wood fire-basin stand

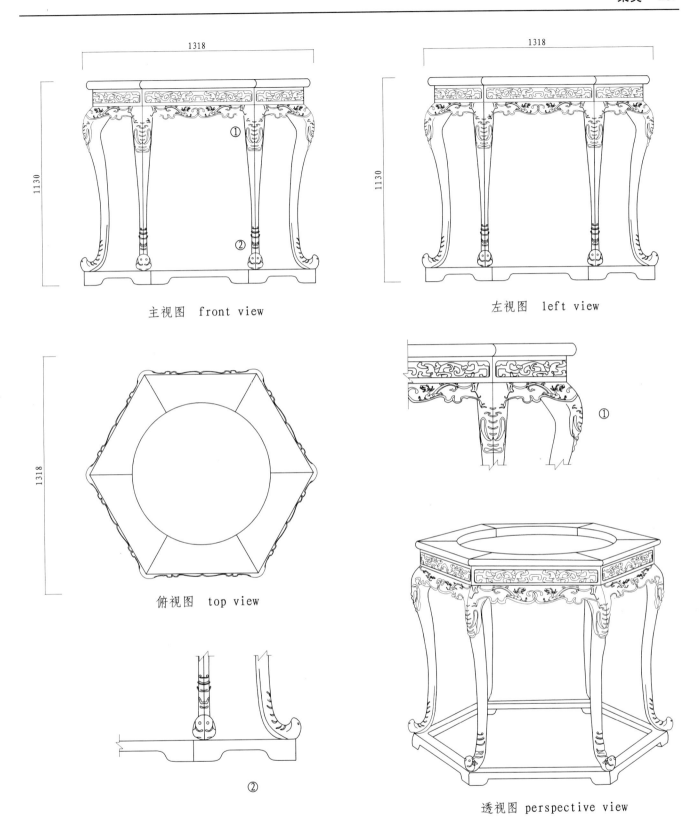

清代黄花梨六方形雕花火盆架

Qing dynasty *huanghuali* wood hexagonal fire-basin stand with ornamental carving

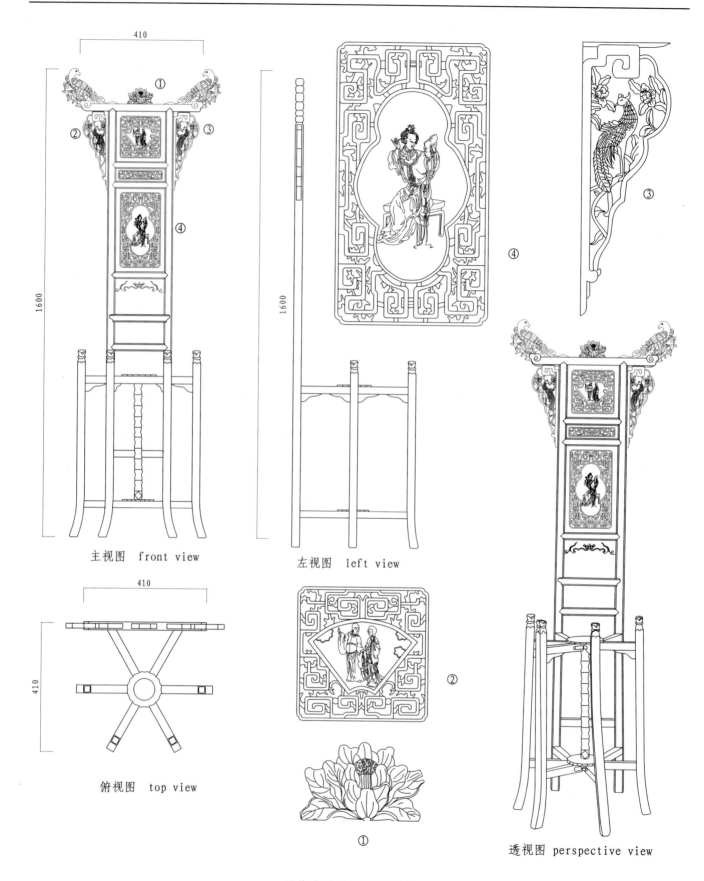

明代宁式花梨木面盆架
Ming dynasty Ningbo style *huali* wood washbasin stand

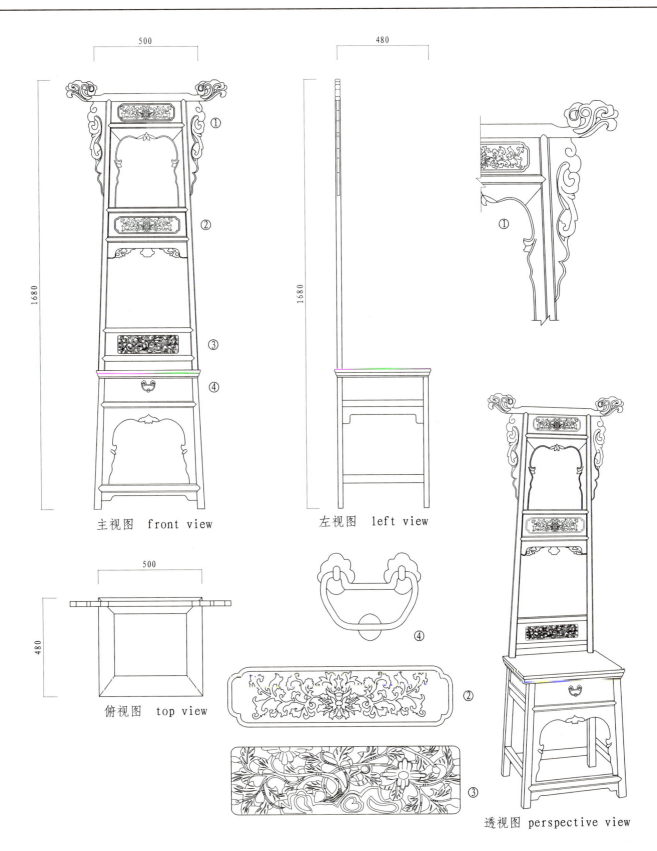

明代宁式花梨木面盆架
Ming dynasty Ningbo style *huali* wood washbasin stand

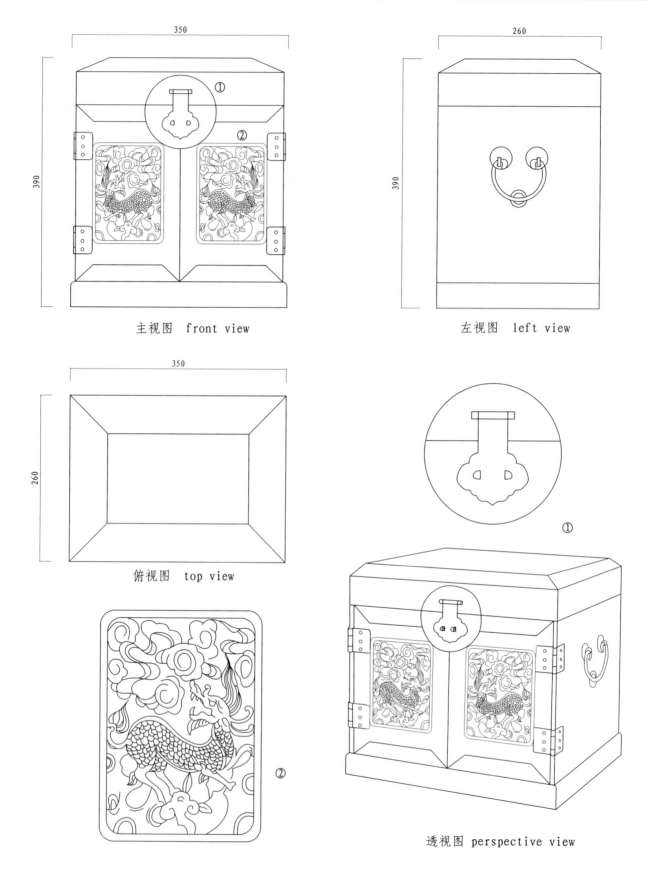

明代透雕门官皮箱
Ming dynasty dressing case with doors of openwork carving

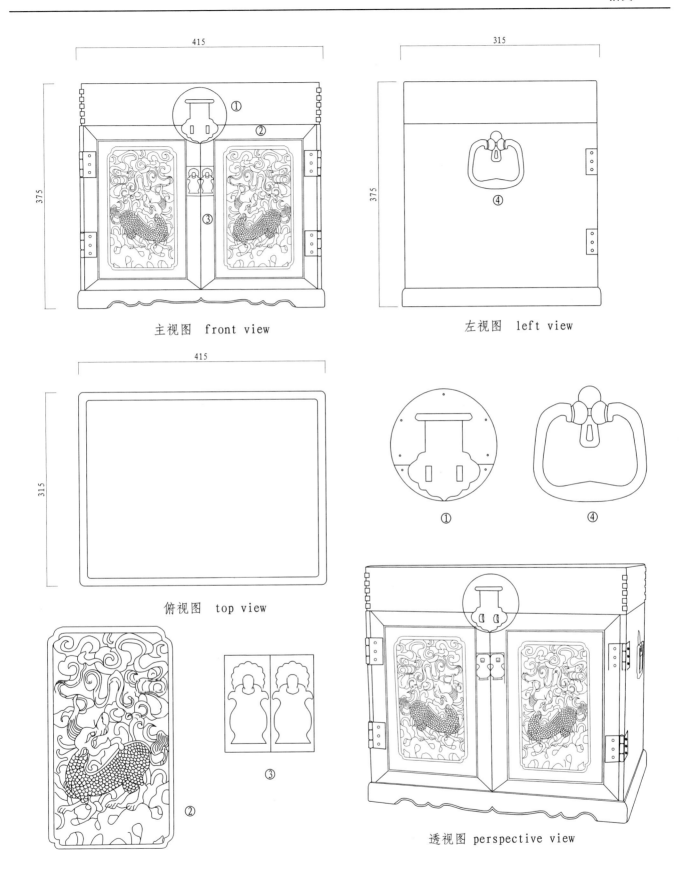

清代黄花梨雕麒麟纹官箱

Qing dynasty *huanghuali* wood dressing case with *qinli* motif carving

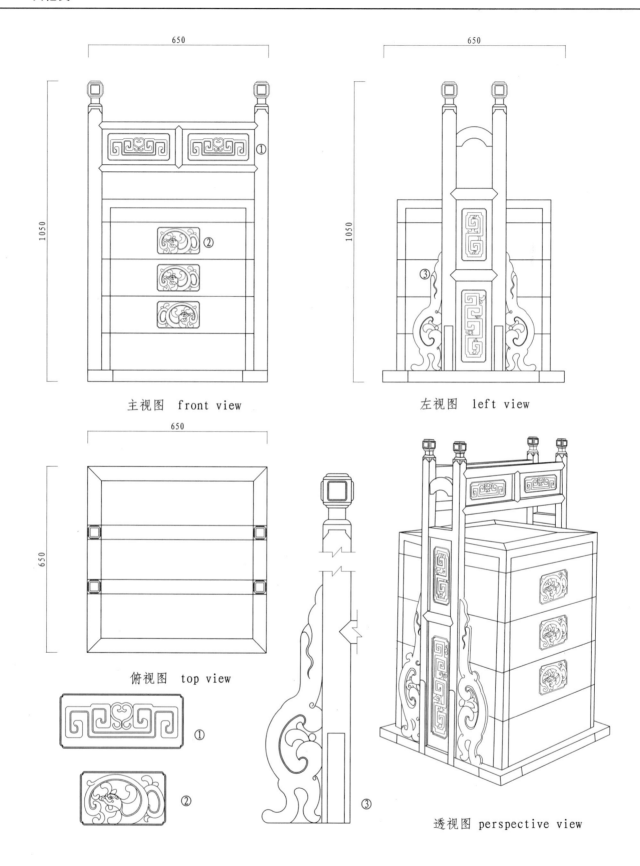

清代宁式榉木扛箱
Qing dynasty Ningbo style *ju* wood box carried on a pole

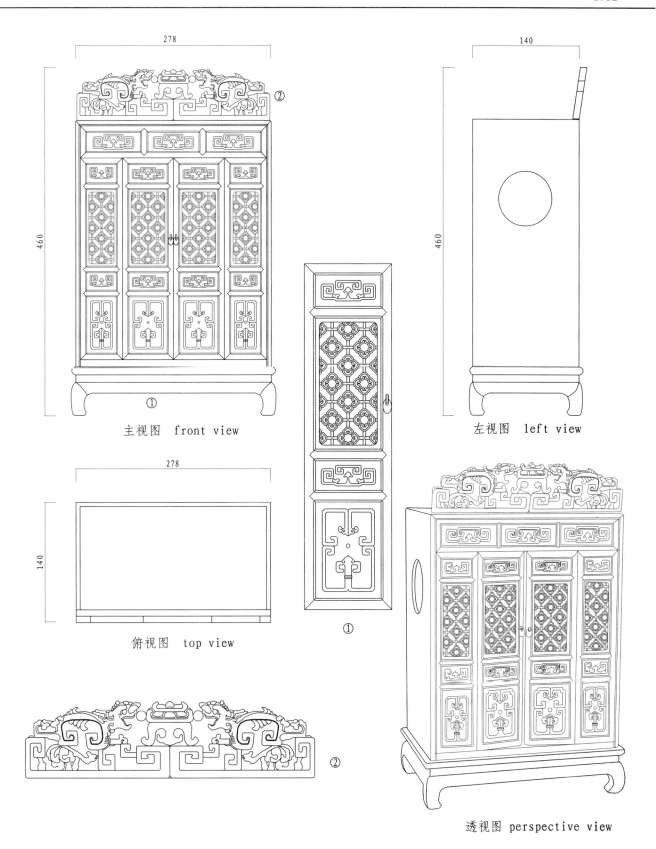

清代紫檀镂空雕花佛龛
Qing dynasty *zitan* wood Buddha chest with ornamental openwork carving

木雕书法　　Calligraphy designs of wood carving

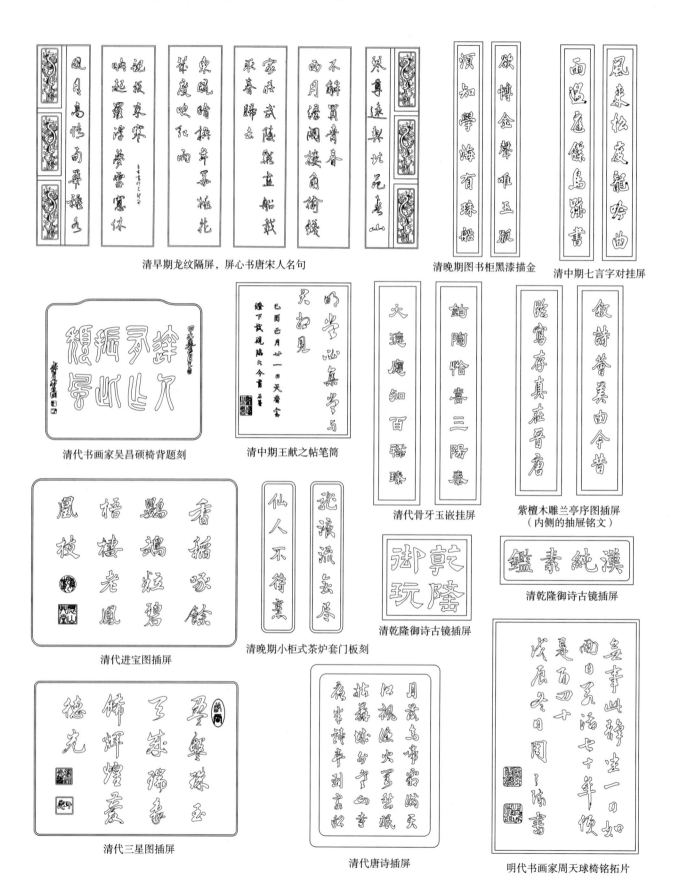

横材和立柱的端头纹样　Finial Designs of horizontal and vertical materials

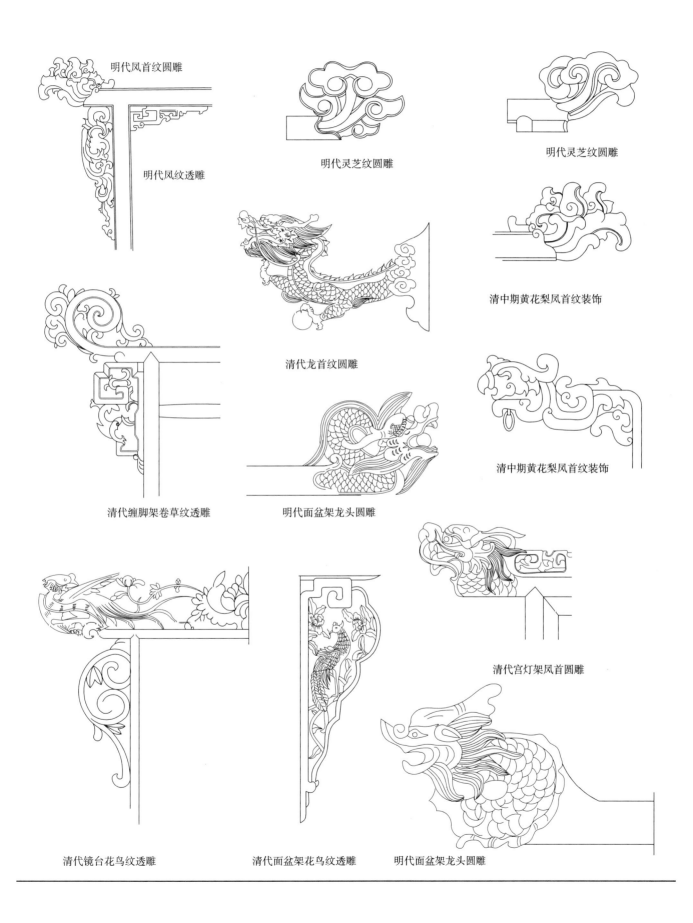

站牙（底座）纹样　Designs of upright brackets and spandrels

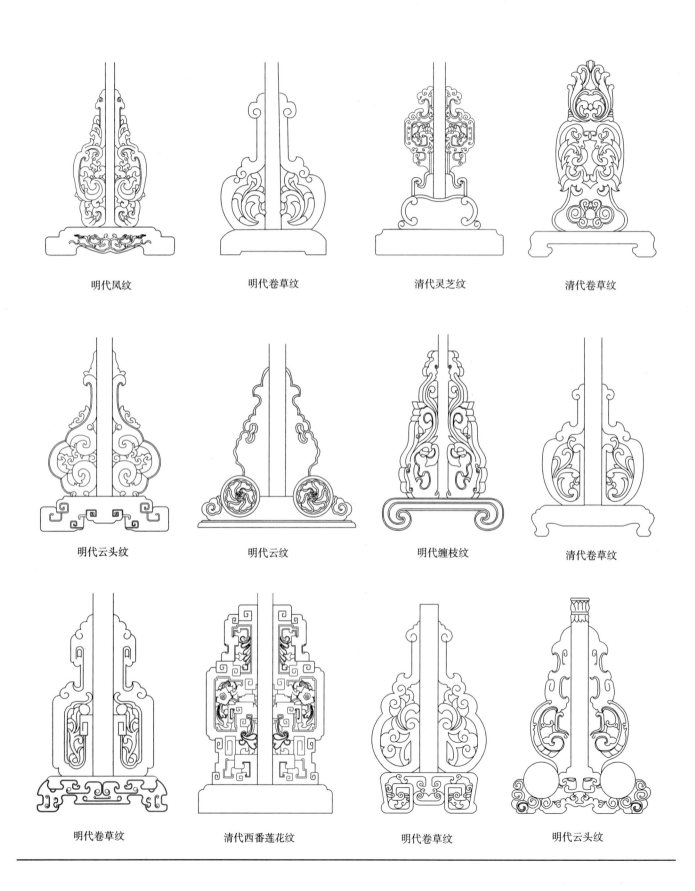

明代凤纹　　　　　明代卷草纹　　　　清代灵芝纹　　　　清代卷草纹

明代云头纹　　　　明代云纹　　　　　明代缠枝纹　　　　清代卷草纹

明代卷草纹　　　　清代西番莲花纹　　明代卷草纹　　　　明代云头纹

角牙纹样　　Spandrels designs

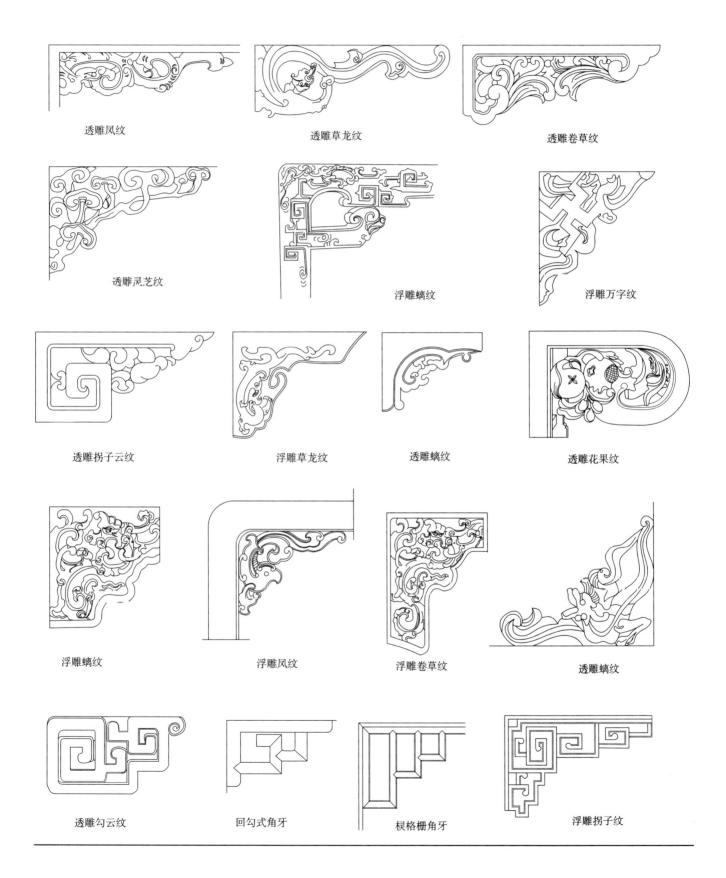

结构与牙头纹样 Designs of structures and spandrels

插肩榫云纹牙头　　夹头榫勾卷纹牙头　　夹头榫云纹牙头

夹肩榫螭纹牙头　　夹头榫拐子纹牙头　　夹头榫螭纹牙头

夹肩榫螭纹牙头　　夹头榫回纹牙头　　平头榫如意纹牙头

夹头榫螭纹牙头　　夹头榫平板牙头　　夹头榫西番莲云纹牙头

结构与牙头纹样　　Designs of structures and spandrels

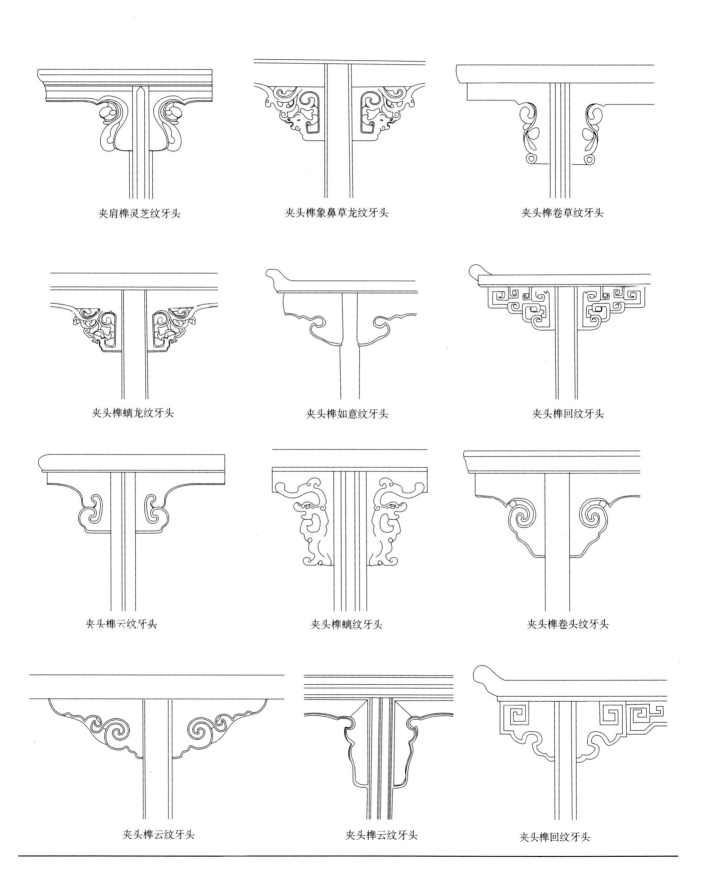

牙板纹样　Designs of aprons

牙板纹样　Designs of aprons

椅背纹样　Designs and motifs on the back of chairs

椅背纹样　Designs and motifs on the back of chairs

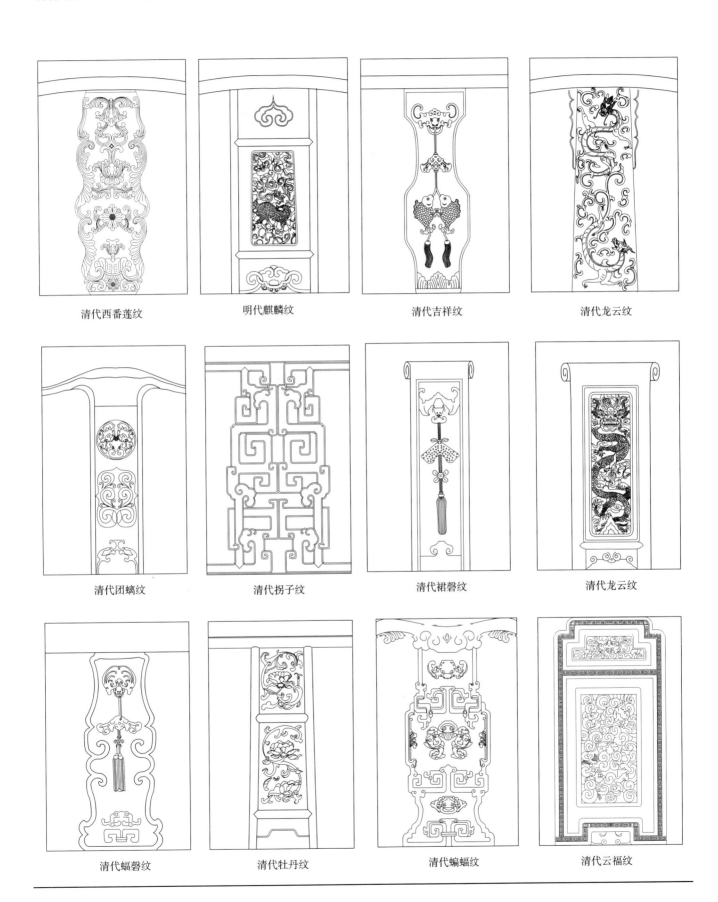

扶手纹样　Designs and motifs on arms of chairs

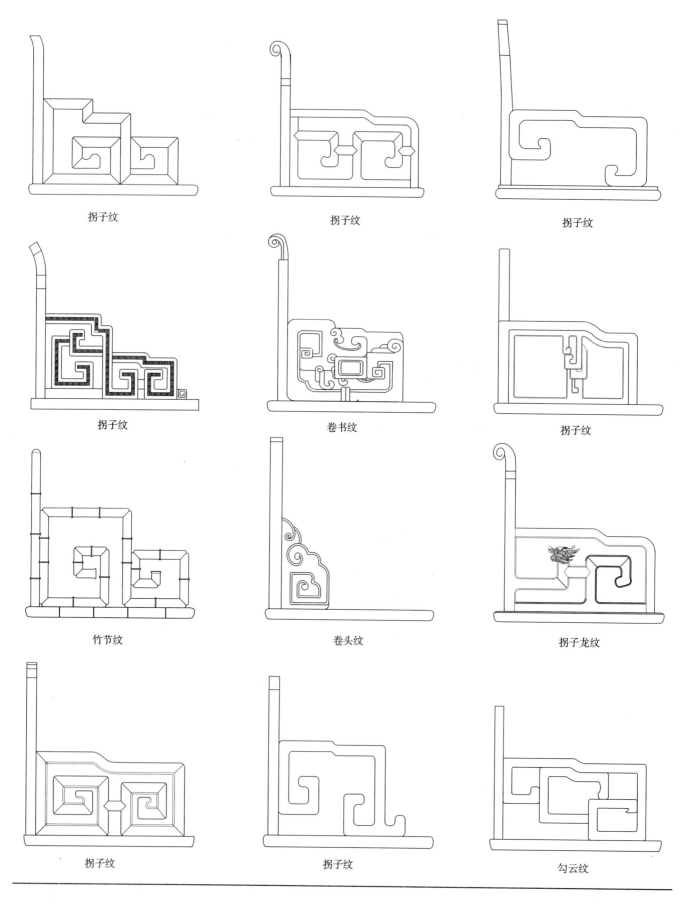

扶手纹样　　Designs and motifs on arms of chairs

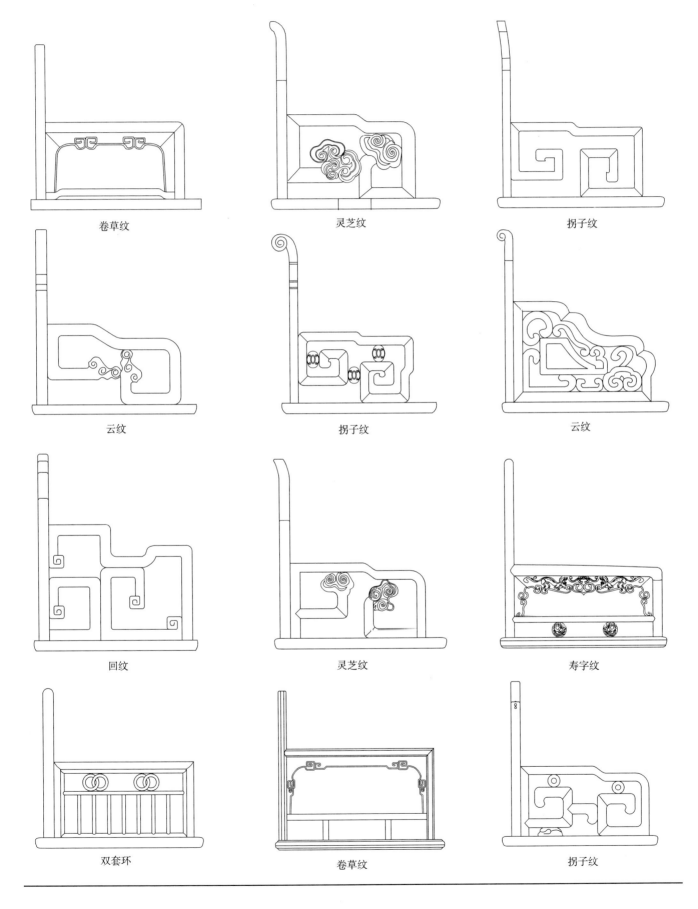

挂檐纹样　Designs of canopy lattice

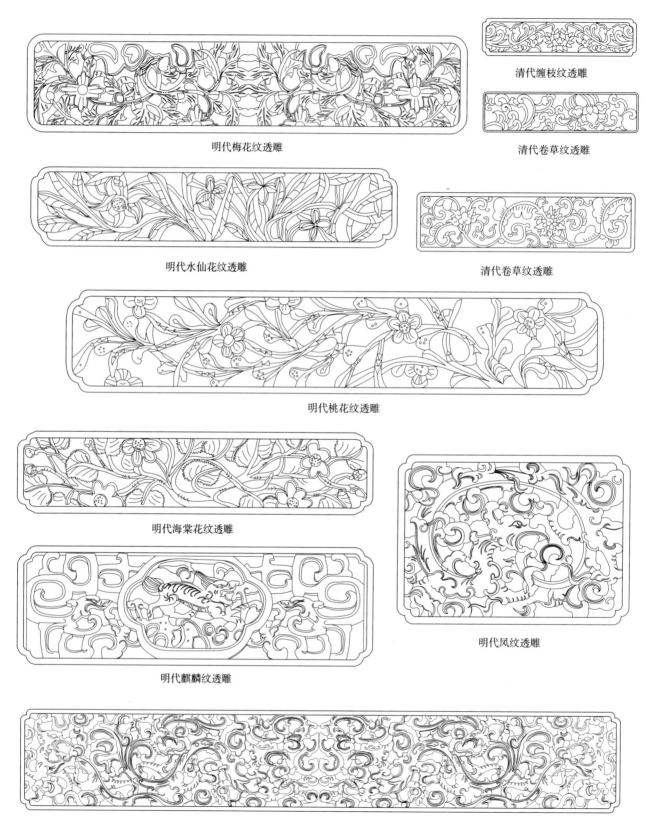

明代梅花纹透雕
清代缠枝纹透雕
清代卷草纹透雕
明代水仙花纹透雕
清代卷草纹透雕
明代桃花纹透雕
明代海棠花纹透雕
明代凤纹透雕
明代麒麟纹透雕
明代螭龙纹透雕

挂檐纹样　Designs of canopy lattice

清代螭龙纹透雕

清代菊花纹透雕

清晚期缠枝葡萄纹透雕

清代云纹浮雕

清代云纹浮雕

清代灵芝纹浮雕

门罩纹样　　Latticework of front part of canopy bed

围栏纹样　Latticeworks of railings

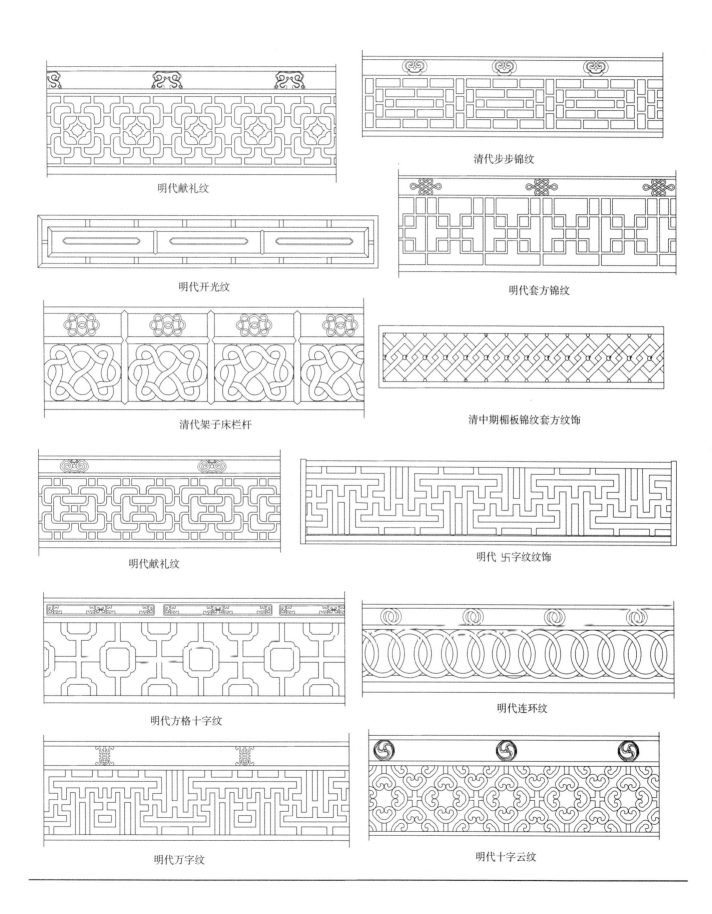

明代献礼纹

明代开光纹

清代架子床栏杆

明代献礼纹

明代方格十字纹

明代万字纹

清代步步锦纹

明代套方锦纹

清中期楣板锦纹套方纹饰

明代卐字纹纹饰

明代连环纹

明代十字云纹

案桌挡板纹样　Designs of large side panels of tables

清代螭纹透雕

明代螭纹透雕

明代灵芝兔石纹透雕

清代花草纹浮雕

清代灵芝双螭纹透雕

明代螭纹透雕

明代螭云纹透雕

案桌挡板纹样　Designs of large side panels of tables

明代麒麟纹透雕

明代花草纹浮雕

明代牡丹花纹浮雕

清代灵芝螭纹透雕

明代凤纹透雕

案桌挡板纹样 Designs of large side panels of tables

凤纹挡板　　　　　灵芝螭纹挡板

凤纹挡板　　　　　草龙纹挡板　　　　平头案腿间挡板"海马负图"纹饰

龙纹挡板　　　　　寿字纹挡板　　　　草龙纹挡板

结子纹样　Decorative mounts

桌案脚样　　Legs styles of tables

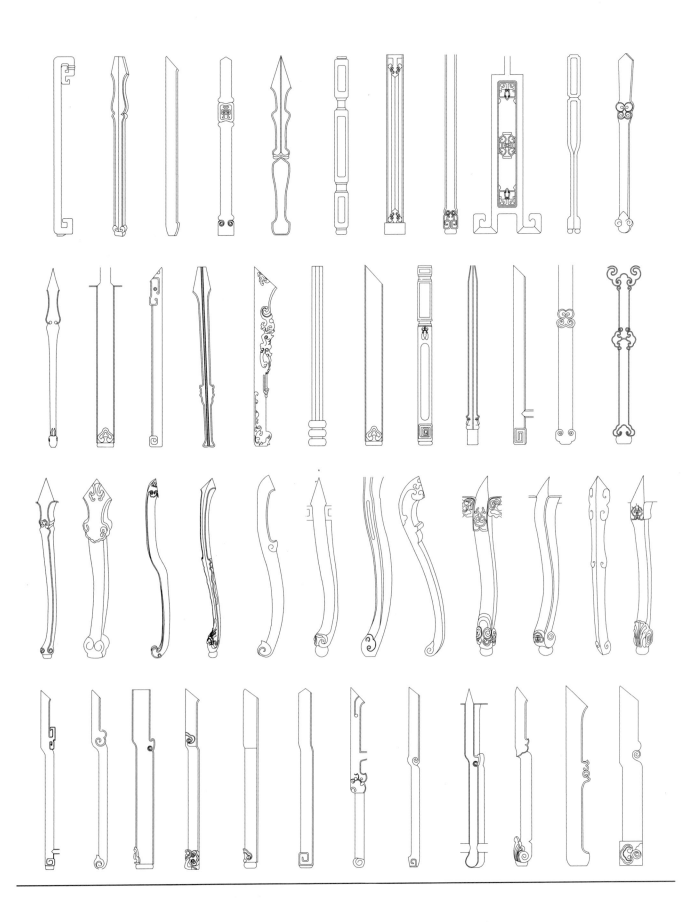

椅凳脚样　　Legs styles of charis and stools

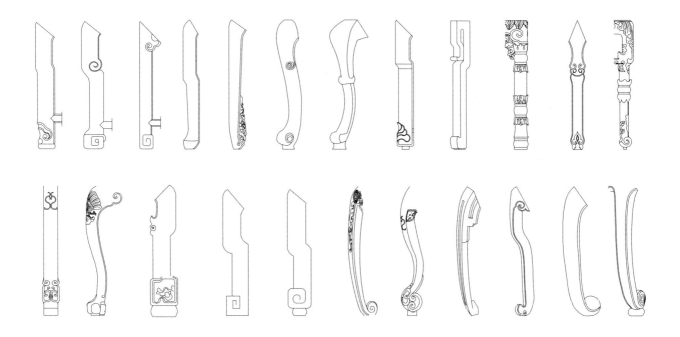

床几脚样　　Legs styles of beds and tables

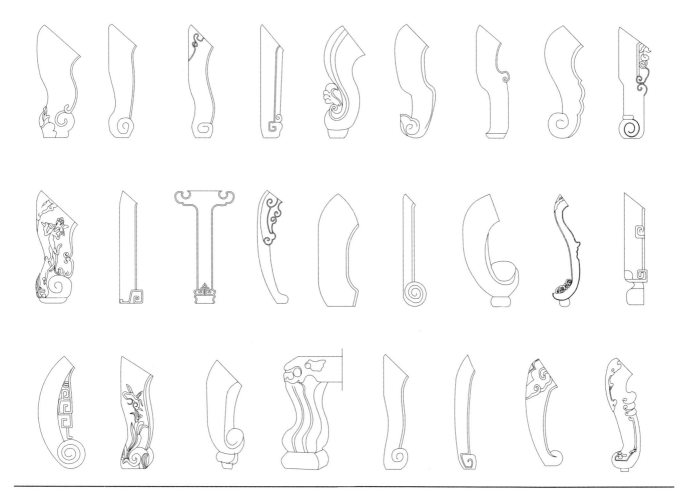

桌角纹样　　Corner and apron motif of tables

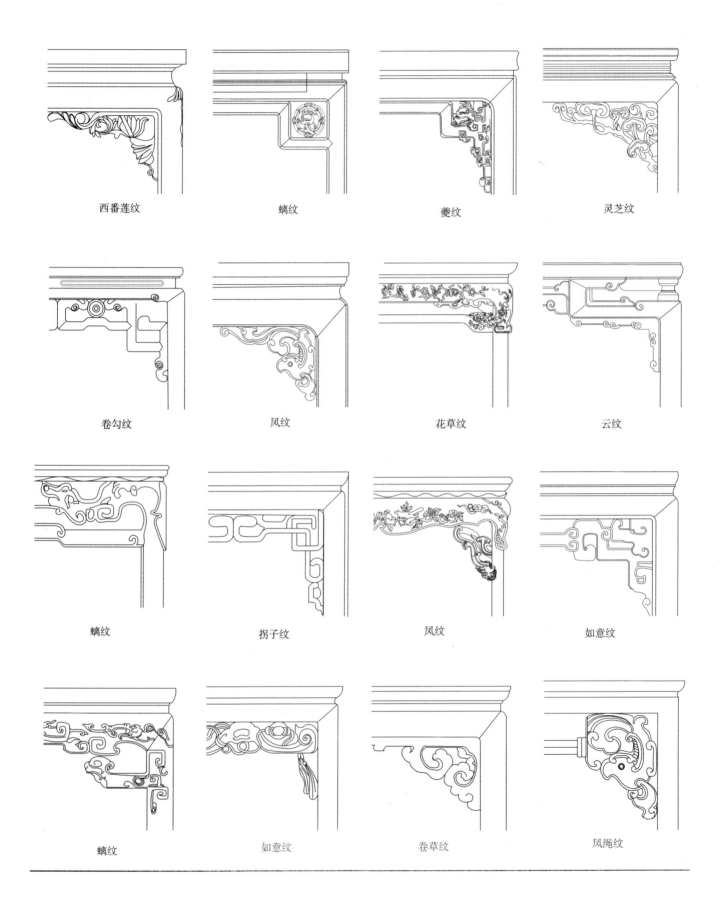

西番莲纹	螭纹	夔纹	灵芝纹
卷勾纹	凤纹	花草纹	云纹
螭纹	拐子纹	凤纹	如意纹
螭纹	如意纹	卷草纹	凤绳纹

托泥纹样　Base stretchers of stand and stool

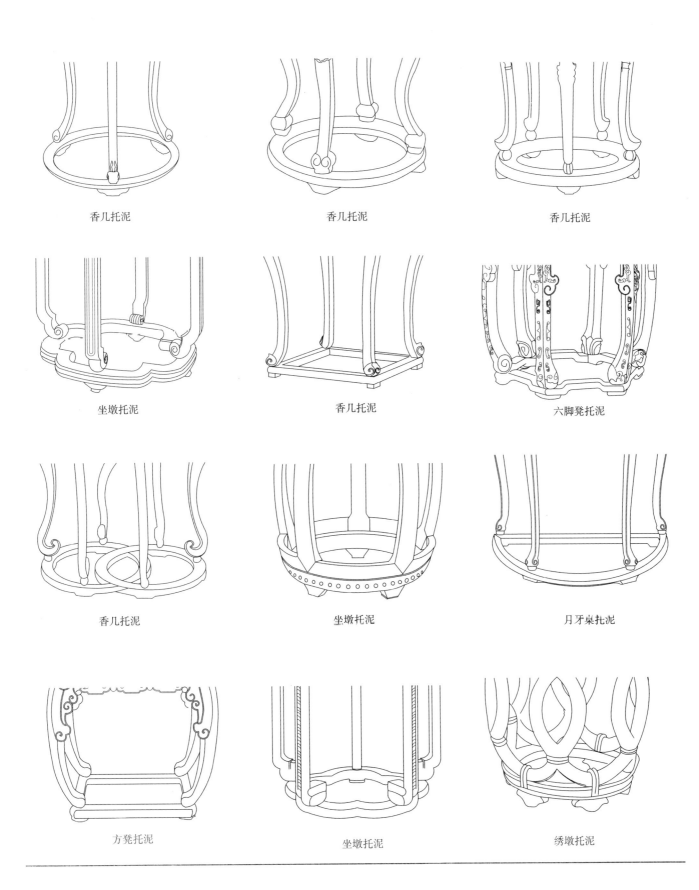

雕花纹样　Carving designs

明末清初　镂空雕竹节边框门绦环板　灵芝纹龙飞图

明末清初　镂空雕竹节边框门绦环板　灵芝纹凤舞图

清代　镂空雕格扇窗花心　狮子舞绣球图

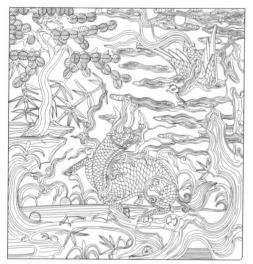

麒麟松凤纹

草龙纹

凤纹

花草纹

雕花纹样　Carving designs

明代麒麟纹透雕

明代花草纹浮雕

明代螭虎纹浮雕

清代草龙戏珠纹透雕

明代四螭禄字纹透雕

明代花鸟纹浮雕

明代麒麟螭虎纹透雕

明代菊花纹浮雕

明代龙喜纹浮雕

雕花纹样　　Carving designs

明代螭虎纹透雕

明代螭虎纹浮雕

明代螭龙捧寿纹透雕

明代螭虎纹浮雕

明代螭纹浮雕

明代福字螭虎纹透雕

明代勾卷纹透雕

明代灵芝纹浮雕

清代平面透雕：博古架、瓷器、书画、棋牌石榴、锦地萍纹

明代虎纹浮雕

雕花纹样　Carving designs

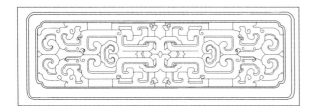

明代云纹透雕

清代螭虎瓜果浮雕

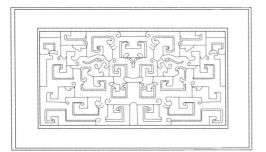

明代凤纹透雕

清代寿字纹样

明代双螭纹浮雕

明代灵芝纹透雕

清代花草纹浮雕

清代花草纹浮雕

明代龙纹浮雕

雕花纹样　Carving designs

明代螭龙纹透雕

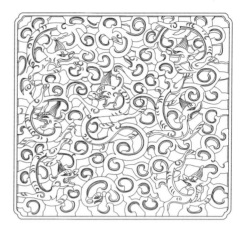

明代螭龙纹透雕

明代螭虎捧寿纹透雕

清代草龙戏珠纹浮雕

明代禄字螭虎纹透雕

明代莲花螭纹透雕

雕花纹样 Carving designs

明代螭龙福中纹浮雕

明代双鱼纹透雕

明代方勾花草纹浮雕

清代草龙戏珠纹透雕

明代螭云纹浮雕

明代凤戏牡丹纹透雕

雕花纹样　Carving designs

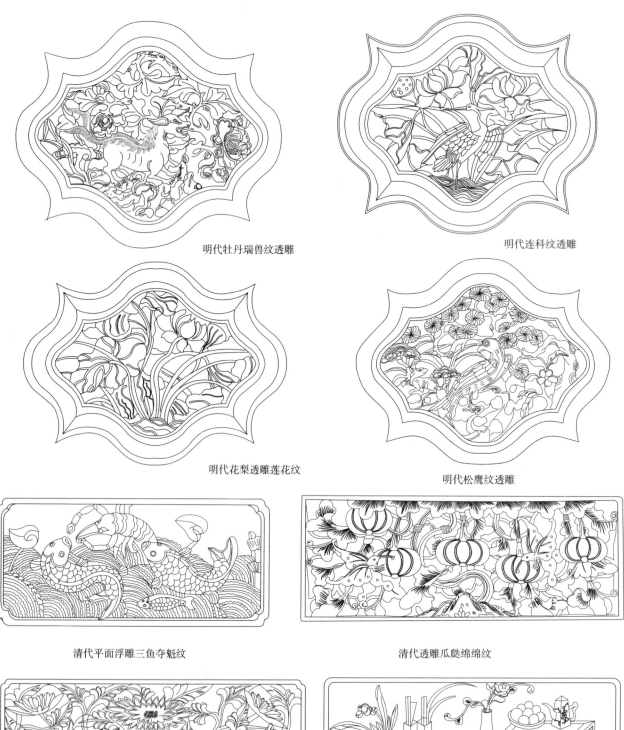

明代牡丹瑞兽纹透雕　　　　　　　明代连科纹透雕

明代花梨透雕莲花纹　　　　　　　明代松鹰纹透雕

清代平面浮雕三鱼夺魁纹　　　　　清代透雕瓜瓞绵绵纹

清代透雕花鸟纹　　　　　　　　　清代平面浮雕博古纹

雕花纹样　　Carving designs

明代螭龙喜字纹透雕

清博古瓶插浮雕

清代寿纹透雕

清代博古平面浮雕

雕花纹样　Carving designs

明代凤纹透雕

明代花草纹浮雕

清代博古瓷杂平面浮雕

明代花鸟纹浮雕

清代博古瓶插纹

清代平面浮雕双英贺寿纹

雕花纹样　Carving designs

明代螭虎戏宝鼎纹透雕

清代平面浅浮雕博古架缠枝莲纹

清代龙云纹透雕

清代博古纹平面浮雕

清代石榴花纹浮雕

雕花纹样　Carving designs

明末花卉纹剔红

清代云龙纹浮雕

清代云螭龙寿字纹浮雕

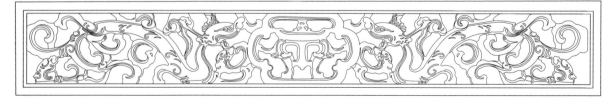

明代象鼻草龙纹透雕

雕花纹样　　Carving designs

清代双龙戏珠纹浮雕

清代螭龙寿字纹透雕

明代花卉凤纹浮雕

清代英雄救美平面浮雕

清代蝙蝠纹寿字纹浮雕

明代瑞兽纹浮雕

雕花纹样　Carving designs

清代博古纹浮雕

清代荷雁纹浮雕

清代凤纹浮雕

清代平面浮雕三羊开泰　　明代牡丹瓶花纹　　明代山水人物纹浮雕

雕花纹样　Carving designs

清代平面浮雕鹰松组合纹

清代博古纹浮雕

清代博古纹浮雕

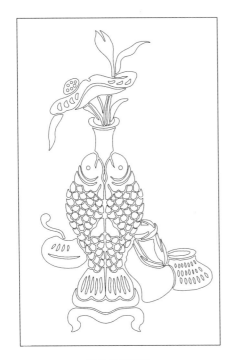

清代鱼纹浮雕

清代花草纹剔红

清代吉祥花草纹

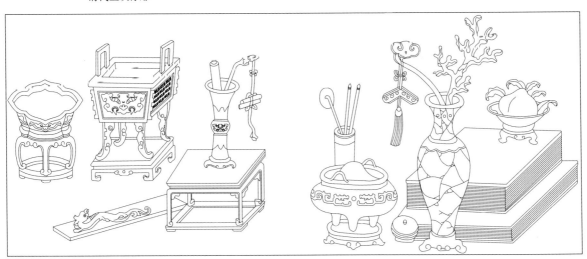

清代博古纹平面浮雕

雕花纹样　Carving designs

雕花纹样（花鸟博古） Carving designs of flowers, birds and numerous antique and precious objects

雕花纹样（祥瑞动物）　Carving designs of lucky animals pattern

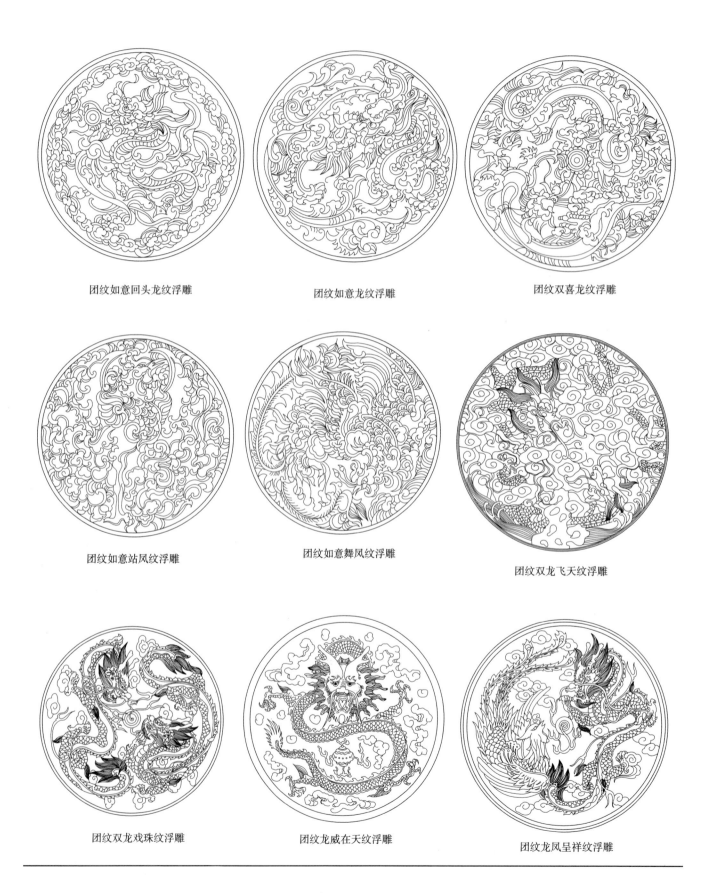

团纹如意回头龙纹浮雕　　团纹如意龙纹浮雕　　团纹双喜龙纹浮雕

团纹如意站凤纹浮雕　　团纹如意舞凤纹浮雕　　团纹双龙飞天纹浮雕

团纹双龙戏珠纹浮雕　　团纹龙威在天纹浮雕　　团纹龙凤呈祥纹浮雕

雕花纹样（八仙图）　　Carving designs of Eignt Immortals Patterns

吕洞宾　　何仙姑　　汉钟离　　曹国舅

蓝采和　　韩湘子　　铁拐李　　张果老

雕花纹样（历史戏曲人物）　　Carving designs of traditional opera figures

太平如意纹　　鹤寿松龄纹　　喜上眉梢纹　　踏雪寻梅纹

麒麟送子纹　　榴开百子纹　　子陵归隐纹　　春牛报喜纹

雕花纹样　Carving designs

镂空雕格扇窗花心　八仙人物图

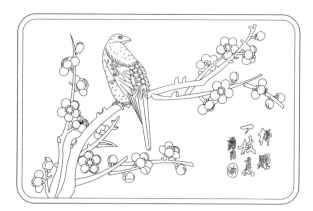

柳赠一枝春　　　　　　　　　　　　　　　　　美人吐绝

明代浮雕门绦环板　梅花鸟兽图　　　　　　　明代浮雕门绦环板　梅花鸟兔图

清代镂空雕窗棂花心　十八罗汉人物图

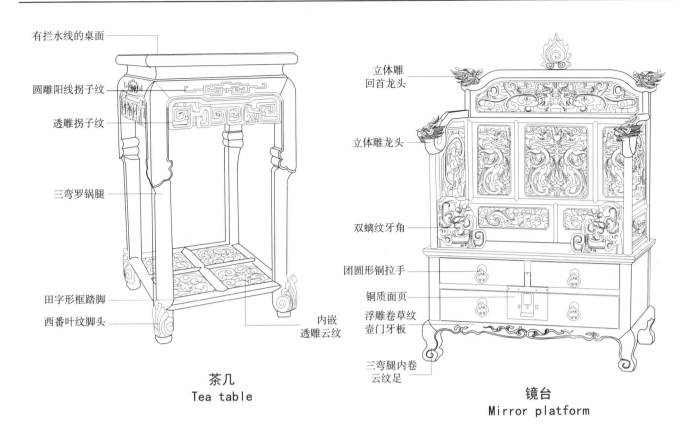

茶几
Tea table

镜台
Mirror platform

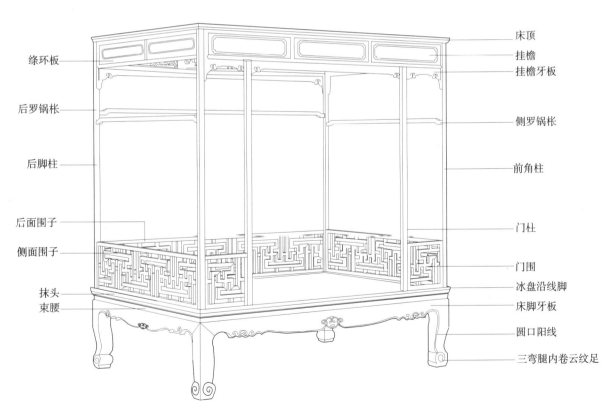

带门围子架子床
Canopy bed with front railings

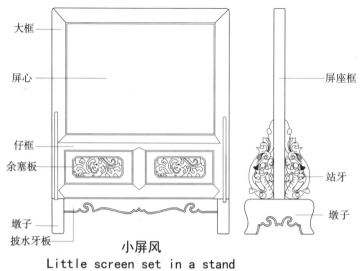

小屏风
Little screen set in a stand

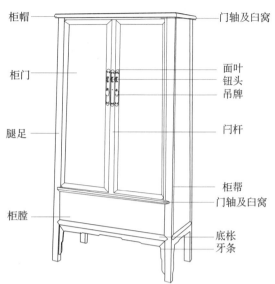

圆角柜
Round-corner cabinet

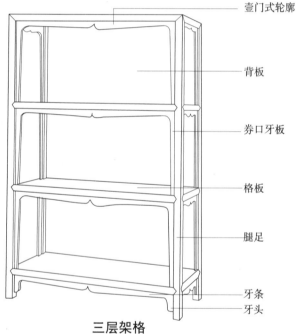

三层架格
Shelf with three shelves

三屉闷户橱（联三橱）
Three-drawer coffer

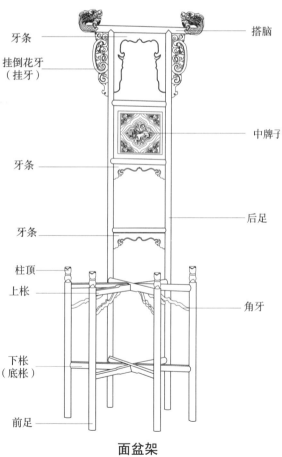

面盆架
Washbasin stand with towel rack